Lessons from the
Living Cell

ABOUT THE AUTHOR

Stephen Rothman, Ph.D., has been an experimental biologist for almost 40 years. He received his training at the University of Pennsylvania in the 1950s and subsequently taught and carried out research as a professor at Harvard Medical School, and since 1971, at the University of California, San Francisco (UCSF). His research has covered a wide range of topics from molecular to whole animal biology and has been published in almost 200 scientific articles in journals such as *Nature* and *Science*. He is the author of *Protein Secretion: A Critical Analysis of the Vesicle Model*. He has lectured around the world about his work, and is perhaps best known for his pioneering studies on the transport of protein molecules across biological membranes. He and his students discovered these processes, which are understood today to be fundamental to and essential for life.

Lessons from the Living Cell

The Limits of Reductionism

Stephen Rothman

McGraw-Hill

New York   Chicago   San Francisco
Lisbon   London   Madrid   Mexico City   Milan
New Delhi   San Juan   Seoul   Singapore
Sydney   Toronto

Library of Congress Cataloging-in-Publication Data

Rothman, S. S. (Stephen S.)
    Lessons from the living cell : the limits of reductionism /
Stephen Rothman.
    p.   cm.
    ISBN 0-07-137820-0
    1. Biology—Philosophy. 2. Cytology—Philosophy. I. Title.

QH331 .R857 2001
571.6'01—dc21
                                                            2001034235

McGraw-Hill

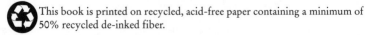

A Division of The **McGraw·Hill** Companies

Copyright ©2002 by Stephen Rothman. All rights reserved. Printed in the United States of America. Except as permitted under the United States Copyright Act of 1976, no part of this publication may be reproduced or distributed in any form or by any means, or stored in a data base or retrieval system, without the prior written permission of the publisher.

1 2 3 4 5 6 7 8 9 0   DOC/DOC   0 7 6 5 4 3 2 1

ISBN: 0-07-137820-0

Printed and bound by R. R. Donnelley & Sons Company.

This book is printed on recycled, acid-free paper containing a minimum of 50% recycled de-inked fiber.

To Bertha and Abraham Rothman,
whom I can never repay.

When I consider every thing that grows
Holds in perfection but a little moment,
That this huge stage preseneth nought but shows
Whereon the stars in secrete influence comment;
When I perceive that men as plants increase,
Cheered and check'd even by the self-same sky;
Vaunt in their youthful sap, at height decrease,
And wear their brave state out of memory;
Then the conceit of this inconstant stay
Sets you most rich in youth before my sight,
Where wasteful time debateth with decay,
To change your day of youth to sullied night;
And, all in war with Time, for love of you,
As he takes from you, I engraft you new.

            William Shakespeare, Sonnet #15

Contents

Preface ix

Acknowledgments xix

1 Beyond the Central Dogma 1
2 What Is It That Makes Something Living? 11
3 The Uncertainties 15
4 The Meanings of Reductionism 19
5 Epistemon and Eudoxus 45
6 The Reductionist Experimental Program 93
7 A Real-Life Parable 101
8 The Lesson 125
9 The Making of a Paradigm 135
10 The Experiments 175
11 To Be Parallel or Nonparallel 217
12 The Tests 241
13 The Call to Authority 273
14 Notes 287

Index 295

Preface

Ring the bells that still can ring.
Forget your perfect offering.
There is a crack in everything.
That's how the light gets in.

Leonard Cohen*

Of Paradigms and Swords

Those who equate science with progress, and those who see it as a great danger to the earth and to themselves, often share a common view of what science is and how it works. In the broadest sense, and whether for good or evil, science is seen as a rational, unbiased, or objective means of gathering and interpreting facts about nature in order to provide a clear and rigorous understanding, and eventually mastery, of its properties.

As any scientist knows and as beginning graduate students find out quickly, this is an idealization. When novice graduate students first learn to carry out and interpret their own experiments, they quickly come face to face with the real world of science. Facts, they find, are often malleable and uncertain, not incontrovertible, and interpreting them is quite often less a matter of applying meticulous reason than

---

*The publishers have generously given permission to use an extended quotation from the following copyrighted work. From "Anthem" by Leonard Cohen, © 1993 Sony/ATV Songs LLC. All rights administered by Sony/ATV Music Publishing, 8 Music Square West, Nashville, TN 37203. All Rights Reserved. Used by permission.

providing an opinion. And when they look beyond their own research to the broader evidence in their particular field, they find the same uncertainty and ambiguity in the work of others. Yet, in spite of such difficulties, despite differences in the experiments scientists perform, what they observe when they perform them, and how they interpret the differences, scientific research does seem to progress in a more or less orthogonal fashion, following what appears to be an orderly path.

Because of this appearance, it is often concluded that scientific progress occurs by the sequential refinement of theories that are modified again and again, as new evidence and reason require, to provide an increasingly accurate description of nature's character. But before long, even beginning students recognize that this also is not an accurate description of the way things happen. Experiments, observations, and their interpretation do not exist in some splendid, hermetically sealed world of reason. They are embedded in and are the consequence of a culture that, although scientific, is replete with biases, prejudices, and, significantly, all sorts of suppositions. Scientists may choose to believe certain results but not others, or to accept a particular interpretation but not another, and not solely because reason demands it of them.

As a community, scientists come to a shared sense of what is known and understood, what remains to be learned, how to learn it, and, finally, what is possible and what is not. They must, or they would never be able to accomplish anything. These collective convictions are taken as an accurate description of our current understanding of nature's properties and as such are passed on to students for their education and preparation.

Thomas Kuhn, in his classic book *The Structure of Scientific Revolutions* (University of Chicago Press, 1962), called such shared beliefs *paradigms,* a word he borrowed from the

Preface

social sciences and that has long since entered our everyday vocabulary. At any given time, he said, experts in a field of science tend to hold a common and comprehensive view of the system under their scrutiny—the paradigm. Kuhn painted a portrait of science as a series of such paradigms (held by different groups of scientists), supported by evidence and reason—often by substantial evidence and for good reason—but also in varying measure by biases, assumptions, suppositions, rationalizations, opinions, prejudices, fallacious reasoning, personal animosities, political considerations, and every other human trait imaginable.

The paradigm represents a shared value judgment about nature, and value judgments are rarely if ever the product of evidence and reason alone. Kuhn realized that new evidence, however compelling, did not necessarily move scientists, like so many yielding reeds in nature's wind, to alter long-held and respected views. Quite to the contrary, contradictory evidence often left older views more deeply entrenched, less open to question.

In Kuhn's view, science was inescapably and unavoidably a social activity, as well as an analytic one. Without perfect means to make the truth manifest and certain—means which usually, if not invariably, do not exist—there is no other option. For some students, coming to appreciate these difficult facts of scientific life can be demoralizing, and many a promising scientific career has ended before it began because of them. How can I proceed in the face of such uncertainty and ambiguity? How can I ever learn the truth if all I can do is seal the holes in my knowledge with shared presumptions about nature?

Kuhn said that most scientists work *within* a particular paradigm. He called this *normal science.* The task of normal science is to seek evidence to bolster the paradigmatic

Preface

view, and to interpret whatever is found in its terms. The bulk of scientific progress occurs in this fashion, as new things are learned and fit to prior belief, proceeding by means of small questions and modest steps. Kuhn argued that major shifts in our understanding of nature, on the other hand, come about only when the foundational tenets of a discipline are seriously challenged: when evidence is sought not to bolster the paradigm, but to dispute it. Such new understanding, he argued, comes about by sea change, by cataclysm, indeed by revolution, not by some smooth and reasoned process of gradualism.

But who would produce the revolution? Who would break the covenant of understanding among a community of knowledgeable experts? After all, that covenant reflects their common perception and hard-earned wisdom. How could questioning at a fundamental level occur in such a situation? Kuhn points to outsiders, individuals who, ignorant of the paradigm and its understandings, unwittingly bring its foundational beliefs into question. Whatever the cause, when this happens—as it gratefully does from time to time—the established order is threatened, and that threat may be seen as no less than a challenge to reason itself. Confronting the menace, supporters of the paradigm defend their position strenuously through research and argument. They attack the contradictory evidence, its interpretation, and sometimes the individuals who present it. And as they do, Kuhn's revolutionaries continue to hurl rocks at the edifice as it is in the process of being fortified, to see how sturdy it really is.

This is science as a kind of war; a war of evidence and ideas, with its tools of logic and reason all right, but also with weapons of passion, prejudice, and power. Even today, years after Kuhn first presented his ideas, his view of science upsets many scientists, not in the least because he places them in the midst of the ordinary, often irrational world of human emotions, not standing apart from its Sturm und Drang as part of a wholly logical and wholly dispassionate enterprise. But of course scientists *are* only human. They are not logic machines, exempt from error, unswayed by base motives, and responsive only to reason's call.

And so it must be. In spite of science's noble goals, its remarkable tools, and its access to powerful means of reasoning, humans practice science much like they practice carpentry or politics—as the flawed creatures they are. It is from this perspective that this book takes a critical look at one of the central beliefs of modern science, especially of modern biology—reductionism. As meant here, *reductionism* is a method of inquiry whereby a system is broken down into its constituent parts and studied at the most fundamental level possible. As we shall see, reductionism has been science's driving force since its emergence, and has been of inestimable, even transcendent, value. It has been far more than a powerful tool. But as with many a good thing, it has been a double-edged sword, with a back edge that is dangerously ragged.

With time, wielders of the sword often lose sight of the difference between its two edges. It is on the shoddy edge that the following narrative focuses. We will be particularly concerned with its impact on our view of life and how we go about learning its secrets. The scientist starts confidently cutting with the blade's sharp edge, but insidiously, and apparently unavoidably, the other edge becomes the tool of choice, as the scientist becomes an unwitting victim of inattention. In the process, the power of evidence and reason is diminished and replaced with crude bias and its constant companion, authority. In what follows, I will explain how and why this occurs.

Preface

A Personal Consideration

I have taught and studied biological systems at various levels of organization—from the molecular, to the cellular, to organs and tissues, and finally in whole animals—for about 40 years. In doing so, I have invariably encountered great difficulty integrating understanding at the lower levels of organization into a broader perspective of the whole. I thought there were two enormous barriers. First, I could not imagine how one could explain the physiological—that is, the functional—wonders of organs and organ systems, like the heart, the circulatory system, the gastrointestinal system, the kidney, and most emphatically the central nervous system, simply by listing every last molecule and every reaction involved, no matter how well we understood those molecules and reactions. It seemed self-evident to me that this was not enough. The material instantiation of biological beings, though unmistakably crucial, was insufficient to explain physiology, and hence to explain life.

Second, and equally important, I could not see how one could discover the mechanisms that underlie various physiological processes without having first established the properties of those processes themselves, whole and intact. It seemed to be widely believed that if we could elucidate the parts of the system well enough, then sooner or later, if we had the necessary skills, insight, and talents, we could understand the mechanisms that underlie any process of interest. This attitude ran counter to everything I knew about physiology. To understand the mechanism, one first had to describe the properties of the phenomenon of interest. Only then could it be grasped. This also seemed self-evident to me. How could one establish the mechanism for a particular process if that process was not known? How would we ever know what to look for?

Preface

This problem became one of central concern in my own research in the mid-1960s. As a newly minted assistant professor at Harvard University, I had developed an interest in how cells secrete their central organic products, proteins. There was already a relatively mature model, part of a larger paradigm, that was widely believed to provide an accurate description of how this occurred. Today this model is known as the *vesicle model* or the *vesicle theory of secretion*. I will describe it later; for the moment it is not important.

When I first began to study protein secretion, I was not well versed in the evidence that supported the vesicle theory, and found it necessary to spend time in the library before I could chart my own path. I had to understand what was known, but also where the lacunae of ignorance lay. What I discovered was a considerable shock to me. Rather than being substantial, the evidence, what there was of it, seemed weak and ephemeral. And worse yet, often what was considered evidence was far more hypothesis than proof. This was not only a great surprise, but raised two worrisome questions. First, why was the vesicle model so widely perceived to describe the actual events of secretion in the absence of convincing proof? And second, given its lack of foundation, how did it come to be so specific and so complex? I anticipated that such a detailed model would be backed by far more convincing evidence than I was able to find.

Equally disconcerting, there seemed to be precious little known about the process it sought to explain. As best I could tell, the model professed to explain a process that was barely understood beyond its mere occurrence. Of course, it was known that secretion occurred, and also that it could be increased or decreased by various stimulants and inhibitors, but that seemed to be about all. This seemed something like sending a rocket to the moon, knowing that the moon existed and knowing its general location (up there and relatively far away), but not knowing how far the rocket would have to travel, in what direction, or to how large a target. It might get there with a great deal of good luck, but you certainly couldn't count on it.

It was with these realizations that my students and I began our experiments. We sought to connect the actual, measurable properties of the process of secretion to the underlying mechanism. As we explored the natural system, the vesicle model served as our guide by providing a framework for predictions. The task was to check the natural process for the properties predicted by the model. If we found them, it would be affirmed. If not, it would have to be rejected or modified.

Rather than salving our concerns, our initial tests raised more questions about the vesicle model than they answered. This spurred us to make further predictions and undertake further trials. But the additional results only served to intensify our uncertainty. It was not long before we began to publish our observations, and wonder aloud (that is, in print) about the goodness of fit of the standard vesicle model to the process it was thought to explain. If secretion did not occur as the vesicle theory proposed, then how did it occur? Along with our questioning, we introduced an alternative hypothesis for the mechanism of secretion. We can call it *direct transport* for the moment.

In reporting this research and commenting on its implications, the die was cast. Quite naively, at least in retrospect, we expected that others would see the weaknesses in the vesicle model as we did, and would join us in further attempts to validate or modify it to more accurately reflect nature. For the most part, this did not happen. Most experts in the area assumed, much to our chagrin, that our observations were erroneous or invalid, however straightforward they seemed to us. Our contradictory evidence was put in doubt, not the vesicle model. And when it eventually became clear that our observations were accurate, it was assumed that they must fit the vesicle model somehow, even if it was not clear how.

In part this was because it was assumed that the alternative we had proposed was not possible. Although this assessment has subsequently been shown to be wrong, at the time it was inconceivable that the case could be otherwise. Since direct transport was the only available alternative, and it was not believed to be a credible choice, the vesicle model had to be correct; and our results, their interpretation, or both had to be in error. This was the opening volley in a controversy that lasted for decades.

Although a discussion of the vesicle model is an important part of this book, I am not concerned with its truth or falsity here. What I *am* concerned with is what I believe to be a flawed—indeed, a fundamentally erroneous—*method* of learning that is exemplified by the familiar experiments that support the vesicle theory, because it cuts across all fields of experimental biology. This method is a consequence of the misuse of reductionism, the ragged edge of its sword. The vesicle model is one of two illustrations of the application of this particular form of reductionism to laboratory research that I will discuss at some length. Let me reiterate so as not to be misunderstood. My concern here is with method, not ultimate meaning. The truth is often found by less than proper means, by serendipity or accident. Bad approaches may, and often do, yield good results, but this does not consequently make them good approaches. And if research is to be scientific, then it must pay attention to its method, and not dismiss it because the truth seems in hand.

Preface

Almost 35 years have passed since I published my first paper describing direct protein transport, and much has changed. There have been a great many papers that support the vesicle model, including the development of an important ancillary hypothesis called the *signal hypothesis.* While much of this evidence has been in the old mold and its interpretation equally dubious, there has been some powerful proof for the presence of certain of the events proposed by the model. Still, when all is said and done, even though almost 50 years have elapsed since it was first proposed, and a great deal of research has been carried out and resources expended, much about the vesicle model remains unsubstantiated.

As for direct transport, the impossible alternative, it is now supported by compelling evidence. Beyond that, we know today that such mechanisms are not only common, not only ubiquitous, but crucial to cellular life. Even so, the vesicle model remains the standard, and I believe most experts still believe it to be sufficient unto itself, though some, including the author, continue to disagree.

Stephen Rothman

Acknowledgments

I would like to express my deep appreciation to all my friends and colleagues who read draft versions of *Lessons from the Living Cell* for their invaluable suggestions and criticism. I would like to particularly mention Lloyd Kozloff, Aaron Lukton, and Paul Silverman for their intelligence and encouragement, as well as my daughter Jennifer for her insightful reading and expert editorial assistance on early drafts of the manuscript. To my wife Doreen, I can offer only my eternal indebtedness for her precious and immeasurable gift of love and support. Also, I cannot write about the vesicle theory and membrane protein transport without recognizing with affection and appreciation all of the students who worked with me on this problem for their earnest commitment and at times unbidden sacrifice. And finally, I would like to express my special gratitude to Amy Murphy for her welcome interest in the book and her inestimable editing skills, as well as to the rest of the editorial staff at McGraw-Hill.

CHAPTER 1

Beyond the Central Dogma

What are little boys made of?
Snips and snails, and puppy dogs' tails;
That's what little boys are made of.

What are little girls made of?
Sugar and spice, and everything nice;
That's what little girls are made of.

Anonymous nursery rhyme

A recent article in *Science* magazine by two well-known biologists, entitled "Genomics: Journey to the Center of Biology," ends by asking whether a full understanding of a "protozoan or peacock" may someday be achieved by "knowing only its DNA sequence?"[1] Though the authors admit that this cannot be done today, they say that they "proceed with optimism" that it will be possible sometime in the future. The central premise of this book is that it will never be possible and that such optimism is misplaced.

---

[1] E. S. Lander and R. A. Weinberg, "Genomics: Journey to the Center of Biology," *Science*, **287**:1777–1782 (2000).

Lessons from the Living Cell
The Human Genome Project and Genomics

If one follows press commentary or popular books in biology and medicine today, much less the scientific literature, one cannot help but get the impression that a full understanding of life lies just over the horizon. Scientists seem to have concluded firmly and clearly that living beings, humans included, can be explained entirely in terms of their material composition—the stuff they are made of—and the chemistry of these substances. Beyond this, and most remarkably, scientists seem to have determined that all of life's consequential aspects can be attributed to a single molecule, the master molecule known as *deoxyribonucleic acid* (DNA).

Accordingly, life can be understood as a complex chemical system at whose center rests an almost imperceptibly small number of these molecules. Just 46 molecules of the DNA polymer are the pivotal ingredients of human life. They carry it from generation to generation and quite literally contain the code of our being—our genes. That code, found in particular variations in the sequence of the four molecular subunits that make up DNA, apparently contains some 30,000 to 35,000 genes, which in turn may provide the information for as many as 100,000 different forms of just one particular class of molecules, the proteins. DNA provides the information necessary for life in its code for building proteins, and proteins, the product of that information and the most complex natural chemical substances known to us, perform life's key functions. If life were a play, DNA would be the playwright, and proteins the actors. Together they are responsible for all of life's salient properties.

And because proteins are chemicals, what they do and how they do it can be explained by general laws of chemistry.

As a consequence, and at a most fundamental level, life can be reduced to—that is, be explained in terms of—basic chemical properties. Life is simply the distinctive application of these chemical properties by particular protein molecules.

This extreme reduction of life to chemistry is the remarkable conclusion of a great body of scientific research in molecular biology and biochemistry conducted during the past century. Its culmination was the Human Genome Project, which sought to specify in all relevant detail the chemical composition of our DNA. It seemed that by realizing this goal, we would gain a comprehensive understanding of life. But the genome has been sequenced, and nothing even remotely approaching the complete mastery of life's character has been achieved. We have learned many things, and will no doubt learn many more, but the key to a comprehensive understanding of life is not one of them.

Such an understanding was never realistically achievable this way. But advocates of the genome project reason that it is just around the corner. They see the sequencing of the genome as the essential prerequisite to such far-reaching comprehension. That will occur when two additional tasks have been completed. First, the exact parts of DNA that carry the code for each of the various protein molecules need to be identified; second, having identified them, we must learn what they do. When this information is in hand, when the structures and functions of all of our protein molecules have been deciphered, we will have described life's nature in its entirety, or at least in all important respects. Fulfilling this ambitious objective is an important goal of an exciting new field called *genomics*. Some believe that however long it may take (the *New York Times* called it the work of 100 years), genomics will follow the critical path laid out by the Human Genome Project and bring the journey to a successful conclusion, resolving once and for all who and what we are in the most intimate chemical terms possible.

The Central Dogma

The equation of life with its materiality, with its DNA and protein molecules, is sometimes called *genetic determinism*. Genetic determinism is the belief that who and what we are is wholly determined by the genes, by the DNA, that we inherit from our parents. In this view, nothing less than life itself is the product of this inheritance. Consequently, it should come as no surprise that the authors of the article cited at the beginning of this chapter began their discussion of genomics with the following statement: "Without doubt, the greatest achievement of biology over the last millennium has been the elucidation of the mechanism of heredity." Elucidating the mechanism of heredity has unquestionably been one of the greatest achievements of science, but is the discovery of DNA and its role in genetics biology's single greatest achievement over the past 1000 years? What about the discovery of the centrality of cells to life–cell theory? What about the understanding that living things obey the same laws of physics and chemistry as inanimate matter? Or the discovery of the function of the heart and circulatory system, or that of the nervous system and its relationship to the mind, or the realization that there is a substance (oxygen) in air that is necessary for animal life—or, most important of all in my view, Darwin's and Wallace's theory of evolution by natural selection? We could easily extend this list, but the point is that as great an accomplishment as genetic theory has been, to see it as the ultimate achievement of biology is to ignore much that is of supreme importance. This is why it is appropriate to call it a *dogma*. As with all dogmas, it ignores or relegates to insignificance important aspects of reality outside its conceptual boundaries.

In this estimation, we are DNA and proteins, and our being is the result of their particular and respective properties. While this characterization leaves out many important details of the story, such as the role of other substances like *ribonucleic acid* (RNA) or parts of DNA that do not encode proteins but allow for their expression, these omissions change nothing. From this perspective, the rest of biology—the rest of what we understand about life, human life not excepted—may be interesting, and worthy of understanding, but it is all commentary, mere footnotes to the main narrative about DNA and protein molecules.

Life beyond the Central Dogma

Yet something seems wrong with this conclusion, and it is not merely that there are other great discoveries to acknowledge. In everyday life and in life's multitudinous expressions, there appears to be a vast chasm between the view of the central dogma and our own experience—more broadly, between the molecules of life, what they are and what they do, and what we, as living creatures, are and do. Genetic determinism says that the latter is simply a consequence of the former. That is, all of life's complex phenomena can ultimately be fully explained in terms of molecules and reactions, namely, reduced to them.

Of course it is one thing to make such a claim, but quite another to prove it. The simple fact is that in most instances we lack any grounded notion of how the gap between life's molecules, chemical reactions, and parts more generally and its broader phenomena is bridged. There are some important exceptions, and they give scientists confidence that similar bridges exist for all of life's features. For example, we have a good—indeed, a penetrating—understanding of how chemical reactions provide us with energy and build our material substance (metabolism), and, as we have been discussing, we understand the chemical nature of inheritance (the expression of genes via protein molecules). But for myriad phenomena of life, the nature of the bridge that connects life's molecules to its intact features is unknown.

We can point to discontinuities between life and its molecules in virtually every area of biology, from the cellular level to the anatomy, physiology, biophysics, evolutionary biology, and ecology of all sorts of organisms, from elephants to flowering trees to lobsters. For example, how does the circulatory system emerge from mere chemicals like DNA and protein? How do our genes shape the heart, provide for its various chambers, for the pumping of blood? How do they give rise to the various vessels that permeate our bodies in different sizes and types in many very particular locations? And, the greatest mystery in my opinion, how do the chemicals of which the brain is composed—including DNA and its genes, but, most important, the molecules that carry information from one nerve cell to another, the so-called neurotransmitters—give rise to its remarkable activities? How genes and their proteins bring about such complexity and organization all by themselves, if indeed they do, remains a mystery in spite of much effort to understand such events in these terms.

Natural Selection and Life

However mysterious such things may seem, they are just the tip of the iceberg. The same mystery is found wherever life's nature seems to transcend its molecular antecedents. We can

Beyond the Central Dogma

get a glimpse of the enormity of the problem by considering what to me is life's central phenomenon: the interaction of organisms with their environment in pursuit of survival. What genes, what DNA, account for or cause this? What gene leads the cheetah to attack a gazelle, or the rabbit to run?

To a removed observer, biology seems to be two wholly different fields with radically different understandings of life's nature. There is *experimental biology*, which seeks to discover life's essence in its DNA, its chemical substances, its microscopic structures, and so forth. And there is *evolutionary biology*, which tries to understand life as Darwin understood it, whole and embedded in its environment. Both groups study the same systems, the same life, the same material objects. Both agree that life's chemical ingredients are the substrate upon which whole living objects are built, and that it is through the interactions of organisms with their environment that life is lived. But otherwise they articulate very different visions. To one, life is found in the properties of molecules and reactions and the parts of living things more generally. To the other, it is found in the circumstances of whole organisms and the factors that govern their reaction to them.

Since the discovery of the genetic code, it has become the common wisdom that this seemingly great disparity is more apparent than real. DNA and its proteins are responsible not only for the way we are as individuals, but for the evolution of species as well. Changes in the structure of DNA produced by mutations, by genetic recombination, and in other ways ultimately give rise to new species. It is these differences in DNA that are responsible for the great variety of life that we see and that has existed in the past. And it is because of variations in DNA that one organism survives and another disappears in the face of a particular environmental challenge. That is to say, survival in response to the forces of nature—natural selection—depends upon our genes. As such, evolution and natural selection are at the most fundamental level the evolution and selection of *genes,* not of organisms. The organism is merely their container. When viewed in this way, the great divide between life's materiality and its nature whole and intact vanishes. Genes are the bridge to life.

But equating genes and evolution in this way is just a mental slight of hand. Genes do not in and of themselves make an organism more or less able to adapt to the environmental exigencies it faces. This ability is an exclusive property of the whole intact organism. The gene is the handmaiden, not the master. Most important, genes and proteins are not in and of themselves responsible for the multitudinous features of life that provide for our adaptive capabilities. Yes, these features depend on the materiality of life, but the fact that certain properties are adaptive cannot be attributed to that materiality. As we shall see, they are transcendent qualities of the organism, and it alone.

While molecules and reactions are necessary for life, and are connected to its intact features in one way or another, the paradigm of genetic determinism claims far more. In its terms, our molecules, their reactions, and the structures they form and are embedded in are not only necessary but are sufficient for life. They *are* life and seamlessly give rise to each and every one of its aspects.

At this critical juncture, on the brink of the Genomic Age, it is important to appreciate that this conclusion is mistaken. We cannot go from life's material substance to life itself without adding anything in the process. Molecules and reactions—indeed, all of the necessary material aspects of living beings—are *not* sufficient in and of themselves to give life expression. There is something else and something more to life that goes beyond its material incarnation. In what follows, I shall explain what this is and why this is a scientific conclusion, not merely a philosophical or religious conviction.

CHAPTER 2

What Is It That Makes Something Living?

We may, of course, strike a balance between what a living organism takes in as nourishment and what it gives out in excretions; but the results would be mere statistics incapable of throwing light on the inmost phenomena of nutrition in living beings. According to a Dutch chemist's phrase, this would be like trying to tell what happens inside a house by watching what goes in by the door and what comes out by the chimney.

Claude Bernard

The central question of biology, what we might call its "prime directive," is this: What is it that makes something living? It was this question and the need to seek answers to it that initially justified the study of biology as a discipline distinct from the rest of the natural sciences. All of biology can be seen as an attempt to answer it.

"Prime directive" or not, biologists today do not commonly consider it a question worth asking. In fact, it would be fair to say that it is avoided as foolish or naïve in light of modern knowledge. Over the past 100 years or so, as science made exceptional progress in describing the physical and chemical nature of cells and living things in general, it became more and more tempting to think of life as merely a particular instantiation of chemical and physical mechanisms, mechanisms whose properties we were uncovering with increasing clarity and depth.

Cell Theory

Reductionism's first major achievement in biology was cell theory, the discipline's founding principle, which was first proposed about 175 years ago. Cell theory equated life with the cell. Though Mathias Schlieden and Johannes Schwann independently proposed it in the 1830s, questions about the nature of biological cells had been incubating from the time that Robert Hooke first described them some 150 years earlier. Subsequent to the proposal of cell theory, through the remainder of the nineteenth century, it underwent an extended period of development in which it was simultaneously modified and validated. The end result was that the nature of living beings could be expressed in a few basic statements:

1. All living systems, from bacterium to human, are composed of cells, either singly or in aggregate.
2. Cells are the living thing itself.
3. Life comes only from preexisting life.
4. This happens by a process of cell division.

In addition, the following corollaries were deduced or implied:

1. Aggregates of cells form tissues and organs.

2. Complex organisms are composed wholly of cells and their various products.
3. Although there is great variation among cells from different sources, they are at base a single type of object.
4. Cells are able to carry out all functions necessary for life.
5. As such, cells are capable of autonomous life.
6. Cells reproduce their own particular kind by cell division.
7. Cells of a given type, situated within the same environment, do the same thing—that is, carry out the same function or functions.

Most of the great accomplishments of modern biology are hard to imagine in the absence of these understandings. The emergence of scientific medicine and modern agriculture, the twin engines that have driven great improvements in the life of humankind, are a consequence of them.

Although the latter seven statements on the list apply quite generally, there are significant exceptions. For example, some biological structures are composed of cells fused together to form larger entities or *syncytia*. Or, copies of a given type of cell may not be functionally identical; there may be distinct clones with different properties. Parasitic organisms are not capable of independent life, but are viewed as living because they are able to reproduce and display other attributes of life. The anucleate red blood cell, certainly living by many standards, and crucial for life, is unable to reproduce itself (these cells are derived from special stem cells, not each other).

Nonetheless, in the wake of cell theory, the answer to the question of what it is that makes something living was *cells*.

The cell was the atom of biology, the fundamental unit that underlay the surface appearance of living things. But it was much more than that. It was the living object itself.

Soon the question became: What is it about *cells* that makes them living? Is there any particular part of the cell that is alive? And if so, where and what is this living substance? As biologists delved deeper, seeking the enigmatic substance, life itself seemed to disappear. As it became possible to describe the living state with greater and greater chemical and physical specificity and detail, its uniqueness, how the animate is distinct from the inanimate world, became less and less clear. In one sense the differences between the two seemed immense. Nothing like the particularities of life were to be found anywhere else. But those differences, though multitudinous, did not appear to be fundamental. Life seemed to be nothing more than a distinctive and inordinately complex chemical and physical occurrence. Even though it seemed that the question of what makes something living could now be answered rather fully by listing the physical and chemical properties of cells and organisms, it also seemed to have become irrelevant.

CHAPTER 3

The Uncertainties

[In referring to the system of Descartes and others, Aristotle's spirit] said that new systems of nature were but new fashions, which would vary in every age; and even those who pretend to demonstrate them from mathematical principles, would flourish but a short period of time, and be out of vogue when that was determined.

Jonathan Swift, *Gulliver's Travels*

Much has been written describing the path of biology's rapid progression over the past 100 years to the final conclusion that the essence of life lies in its molecules and reactions, especially in the dance between DNA and protein. In this view, the question of the nature of life is for all intents and purposes a solved scientific problem. Though there is admittedly still much to learn in terms of detail and comprehensiveness, from this perspective life can be completely described by enumerating the causative genes and product proteins.

This is the most dramatic and all-encompassing expression of another, even more far-reaching claim, which is not only central to modern biological thought and research, but exemplifies the modern approach. This is the contention that if we possess sufficient knowledge of the material elements that make up living things—that is, if we possess sufficient knowledge of their parts, such as DNA and proteins—we can describe any and all of their aspects completely. Whether the parts account for life itself, as with DNA and protein, or whether they merely account for a particular aspect of it, such as the contraction of muscle or the division of cells, the claim is that if we hold comprehensive knowledge of the parts, we can fully describe the various processes of life they serve.

Though this belief is widely held, to some scientists it is not merely a mistaken conviction, but one that may lead to seriously misleading conclusions. To the skeptics, biology, like physics, has encountered deep problems and significant unanswered questions as its reach has been extended—problems and questions that this misguided belief has served to conceal. They ask whether knowledge of the parts of living systems, however encyclopedic and rigorous, is sufficient to explain the functioning whole. And they ask in addition whether the inability to make such inferences from the parts to the whole is a fundamental limitation of the modern experimental approach in biology. These questions concern nothing less than the logical and philosophical foundations of modern biology. They pertain to its prime directive about the nature of life, as well as the practical issue of how to go about exploring life systems in the laboratory.

Scientists who are fully satisfied with the path of modern biology might ask, why focus on this? It does not matter whether this premise about the relationship between the parts and the whole is rigorously true. It is true enough for our purposes. They might admit that not all of our knowledge has been obtained in the most desirable fashion, and that our inferences from the parts to the whole cannot always be justified in logical terms; in fact, they are often incorrect. But so what? Scientific research is a pragmatic approach to problem solving. It is by trial and error that we home in on the truth of the matter. Furthermore, whatever its flaws, the modern approach is both time proven and time honored. We have been able to proceed far along the trajectory toward a comprehensive understanding of the living state in this way. Why shouldn't this progress continue?

In response, a skeptical biologist might reply that simple acquiescence to practical accomplishment in a field that considers itself a science would be unfortunate. Science aspires to the deepest understanding. And although common sense and practical experience have been and will no doubt continue to be engines of humanity's progress in science as in life more generally, these avenues of approach occasionally fail us—sometimes in astounding and important ways. Historical example shows that a complete and accurate understanding of natural phenomena cannot arise from common sense and practical experience alone, except as chance would have it—not even from the uncommon common sense and perceptive practical experience of accomplished scientists. In the end, we must strive to apply the most meticulous logic and reason that our human abilities allow. And to do this, we must continuously confront our assumptions and our methods with critical testing and analysis. The nature of the inferences we make is central to any such analysis, and we must examine their character constantly. Only in this way can we judge whether particular scientific knowledge is rigorously founded, or just a matter of opinion.

In science, our knowledge is only as strong as the weakest link in the chain of understanding. Whether we consider questions raised by statistical mechanics, relativity, the wave-particle duality, or the missing mass in the universe, it is the unanswered question, the splinter that sticks in our craw, that leads to new awareness and new knowledge. If we fail to confront such uncertainties and unanswered questions with a critical eye toward our assumptions and the character of our inferences, if we shrink from disputing what we consider axiomatic, we will inevitably be condemned to the unsubstantiated belief that at any given time, we, in essence, understand all that there is to know in a particular field. If we follow such a course, we inescapably limit ourselves to filling in the few missing words in an almost completed crossword puzzle, never questioning the nature of the puzzle itself.

CHAPTER 4

The Meanings of Reductionism

Ask a scientist what he conceives the scientific method to be, and he will adopt an expression that is at once solemn and shifty eyed: solemn because he feels he ought to declare an opinion; shifty eyed because he is wondering how to conceal the fact that he has no opinion to declare.

Peter Medawar, as cited by Theodore Schick, Jr.

Most biologists today call themselves *reductionists*, even if they reject the idea that this is tantamount to holding a philosophical belief. Reductionism is not a single idea or proposition, but a group of closely related (and sometimes not so closely related) abstractions that are often confused with each other. And while I believe it to be unquestionably true that many of the accomplishments of modern biology can be attributed to reductionism, it is appropriate to ask to which reductionism I am referring. The reduction of what to what?

Lessons from the Living Cell

Reductionism as Science

In the broadest sense, science and science studied from the reductionist perspective have been synonymous since the time of Newton. From this standpoint, to call yourself a *reductionist* is simply to say that you are a scientist, and to talk of a "reductionist science" is redundant, or of a "nonreductionist science" oxymoronic. Thus it should not be surprising that those who have questioned the reductionist philosophy and research program have from time to time been viewed as "antiscientific," even if they are scientists themselves.

In biology such skeptics have traditionally been greeted with the charge that they are *vitalists*. Vitalism is the belief that life systems have unique features that follow laws different from those that apply to normal matter. It is those special features that make them living. Early in the development of modern biology, during the middle to later decades of the nineteenth century, there was a great, and seemingly final, debate about vitalism and the special nature of biological matter. The outcome was that vitalism was rejected in no uncertain terms. There was no indication of a special living matter that followed its own rules, and certainly no evidence of a living incorporeal substance as had been proposed for millennia. As more and more became known about the physical and chemical properties of cells, such elusive and seemingly illusory matter seemed unnecessary. It became clear that the chemistry and physics of life was at base the same as chemistry and physics in the inanimate realm. This was a reductionist conclusion; hence, to reject reductionism was to affirm vitalism, or so it was often assumed. Vitalism was rejected with such force that to be called a *vitalist* was the most opprobrious of labels. It designated an individual as being ignorant of modern science and as holding antiscientific attitudes and even magical beliefs.

Defining Reductionism

Confusion about the meaning of the term *reductionism* has been so prevalent that it would come as no surprise if two scientists talking about reductionism thought that they were discussing the same concept, when in fact they were articulating two very different visions. Not only that, but scientists who call themselves *reductionists* do not necessarily ascribe to all reductionist ideas, or for that matter may not even be fully aware of them. Reductionism has been important to the ideas of many major philosophers and natural scientists over the centuries, and comprehensively exploring these ideas, with their roots and permutations, would require a substantial book devoted solely to this subject. My interest is merely in providing clear definitions and describing different ways the terms *reduction* and *reductionism* have been used by scientists and philosophers in modern times. Some of the meanings given by scientists may seem inapt from a philosopher's point of view, and perhaps vice versa, but their appropriateness or lack thereof is not the issue. These uses simply exemplify diverse attitudes about biology, and science more generally, that are subsumed under the rubric of reductionism.

The Origins of Reductionism

Its equation with science aside, reductionism is both a philosophical judgment about nature and a research program for studying it. Even though the research program is based on the philosophy, it is distinct, as we shall see presently. Central aspects of the notion originated with the Greeks, and are usually associated with the great thinker Democritus.

It was Democritus who proposed that material things are made of unseen atoms. This idea was reductionist in three different, though related, ways. First, it proposed that an unseen reality exists beyond surface appearances. Second, it proposed that material objects have an underlying structure to which their overt structure can be attributed. In other words, the surface structure can be reduced to that of the underlying atoms. And third, it implied that this underlying structure is more fundamental than the surface structure and hence provides a deeper understanding of reality. Thus, Democritus' reductionism proposed the existence of a fundamental underlying structure for matter, an unseen reality that we cannot directly perceive.

Modern ideas of reductionism emerged as the result of two of the most important conceptual shifts in human thought; shifts that are often viewed as the shared foundations of modern science. The first is embodied in Newton's laws describing the properties of bodies in motion in a single unified and mathematically rigorous way. With the discovery of the gas laws, which applied Newton's ideas to the microscopic (atomic) realm, all bodies, all material objects, whatever their particular instantiation, were understood to move in a predictable fashion in response to the application of specified forces in accordance with the same universal physical laws.

The second development was Descartes's monumental decision that the body and the immortal soul should be considered separately. This allowed the bodies of both humans and beasts to be examined in wholly physical terms for the first time. Descartes saw the human body as a machine, much like a child's mechanical toy. It was from this perspective that

The Meanings of Reductionism

he proposed that all living things, their soulful nature excepted, are made of ordinary matter. In his view, living bodies were the same as inanimate objects except in the details of their incarnations. As a consequence, they obeyed the same laws.[1]

By the early twentieth century these ideas formed the basis of most scientific research. From Democritus and the Greeks we gained the idea that the natural world contains an unseen, underlying structure that is more fundamental than the surface realities. From Newton, Descartes, and their contemporaries we learned that there are comprehensive natural laws that are common to all types of matter. And finally, Descartes allowed us to look at life systems as being one with the rest of the natural world, not as something apart. These themes form the foundation of both modern science and reductionism, hence their seeming equivalence.

The Unseen

In modern times an important caveat has been appended to Democritus' original reductive notion of a fundamental structure underlying directly observable reality. This warning concerns how we learn about the unseen elements. It cautions us that what we obtain from scientific instruments is not direct knowledge of the unseen. In spite of the common perception that our sophisticated machinery provides direct access to structures that we cannot engage with our primary senses, this is not true. We gain such knowledge only indirectly by deduction. That is to say, deductive reasoning is a

---

[1] In the minds of many scientists and philosophers today, Descartes's exception for the soul has been revoked. As we have learned about the structure and function of the nervous system, soul and mind have joined the rest of the body as just another part of the human as machine. The last word on this often unfriendly, even hostile, merger has yet to be spoken, and in my view probably never will.

necessary prerequisite for explicating the underlying or hidden properties of nature. And as a consequence, it is only through the intervention of the human faculty of reason that we are able to gain such knowledge.

Our machines are predicated on particular understandings about nature, specific scientific knowledge and assumptions. As a consequence, they are only as useful as those understandings are correct. For example, our microscopes are only as reliable in providing us with an accurate description of the microscopic world as our comprehension of physical optics is rigorous. This means that it is not adequate simply to evaluate the physical incarnation of a device, such as how well the lens is made. We must also consider the reasoning that went into its construction. Unfortunately, not all scientific devices and the methods they serve are built on a foundation as secure as physical optics. Our evaluation of the outcome of observations derived from such devices and methods must include a demanding appraisal of their theoretical and technical foundation. Woefully, it is this appraisal that often removes the subject from comprehension by all but the expert.

Universalism

What I call *universalism* is probably the most widely understood meaning of the term *reductionism*, and is adhered to by virtually all scientists in modern times. It simply says that all phenomena in the natural world can be explained in terms of—that is, can be reduced to—certain fundamental laws of nature. As such, universalism is closely related to *natural law*, which states that nature is not random or chaotic, but displays order that humans can interpret. As applied to biological systems, universalism is expressed in the Cartesian proposal that living organisms obey the same laws as the rest of the material world, and can therefore be explained in terms of those laws. It is hard to gainsay this view today. Even if we could imagine new discoveries in *biology* that could lead to a new depth of understanding of the *physical world*, it would seem that laws so derived would be no less a part of that world than others. That is, they would still be general laws of nature, even if they found their most developed expression in living organisms. In my view, it is to the concept of universalism that the successes of reductionist biology can in great part be attributed. To many this *is* reductionism, but the term has accumulated many other meanings, some of which are not so easily justified.

Generalization

If reductionist science is to be universal, then it must also seek to generalize; that is, it must attempt to explain phenomena in the natural world in the most comprehensive terms possible. Generalization is closely linked to universalism, and is sometimes confused with it. But to generalize, an explanation need not be reductionistic, or even scientific. Scientists seek to generalize from observation by the construction of theories. In this view, the better theory is the one that allows for the greater generalization; that is, it applies more broadly to phenomena in the natural world. Ideally, a successful theory should account for all observable phenomena in its area of interest. As such, it seeks to be "universal" in its sphere of explanation.

The attempt to generalize from specific observations is central to the scientific method. It is also a basic characteristic of human thought. We do not simply observe the world. We continually generalize from specific experiences in our daily lives. We make our own personal generalizations, and they are sometimes good ones (those that help us) and sometimes bad ones (those that hinder us). If we have a bad experience in a restaurant, we may decide not to return on the off chance that we will have another unpleasant experience. If we see a tall animal and a small animal, we naturally consider whether they are parent and child, or of two different species. And on and on and on. Not to generalize from specific observations is unthinkable for humans, and probably for many other species as well.

Hierarchical Universalism

There is another, more explicit universalism, that we can call *hierarchical universalism*. It proposes that the various scientific disciplines are hierarchical in nature, from the most fundamental (simplest) to the least fundamental (most complex) in the following order: mathematics, physics, chemistry, biology, and the social sciences. Information from less fundamental disciplines can be explained in terms of or reduced to the rules of the more fundamental ones, but not the other way around. For example, however difficult, social organization can at least theoretically be explained in terms of the properties of atoms, but the properties of atoms can never be explained in terms of social organization.

One may accept the general notion of universalism, but reject hierarchical universalism. For example, we might ask whether hierarchical reductionism is truly universal—that is, applicable in all cases. Certainly, the example of social structure and atoms that I just gave might give one some pause to wonder whether such a reduction is really possible, even in theory. In addition, can we be sure that certain aspects of biology are not more fundamental than particular concepts in physics, or even all of physics? It has been argued that this cannot be the

The Meanings of Reductionism

case because living systems make up a trivial portion of our planet's mass, and an infinitesimal part of the matter in the universe. It does not seem reasonable that such a trivial proportion of known matter would embody more basic laws than the far, far greater mass of inanimate material. Similarly, it can be argued that biological life has been present for only a fraction of the existence of the universe. In these terms, it is a fleeting presence that will disappear as quickly as it appeared. Why would objects whose presence is, relatively speaking, short lived entail rules that are more fundamental than those that apply to more enduring matter?

But some scientists have questioned this conclusion. Even though the proposition that the trivial mass and seemingly fleeting presence of life forms indicate their less fundamental nature is certainly a reasonable one, this does not necessarily make it true. Today there is increasing interest in seeking fundamental rules for the complex organization of systems. These rules would be based on the *relationships* between things, not their underlying *substance*. If such rules can be found, and applied to nature in a most fundamental and general fashion, then the hierarchy from simple to complex, from fundamental to less fundamental, might be turned on its head. This case cannot be convincingly made today; even though there has been much progress during the past 50 years in understanding the notion of complexity and the related ideas of organization, information, content, context, and relationships, our comprehension is still very much in the formative stages.

Grand Universalism

When the concepts of generalization and universalism are taken together and to their logical extreme, they give us the most all-encompassing of reductionisms. This view, *grand universalism,* is of great potential significance, but equally great uncertainty. Grand universalism proposes that, at least in theory, we should be able to explain everything, all of the natural world, with a single, most fundamental understanding that is at once both completely general and totally comprehensive—an algorithm to which all other understanding can be reduced. Some physicists, most famously Albert Einstein, and mathematicians have worked hard to unify all of the known physical forces of the universe into a grand unified theory, or a theory of everything. If the known forces of the universe are both basic and comprehensive, then it is reasonable to expect that a theory could be developed that would explain everything, at least everything physical.

However, even though science attempts to provide the broadest possible umbrella for understanding and attempts to reduce phenomena to the most universal principles possible, the realization of a single complete explanation for all phenomena in the physical world is quite a tall order. Some mathematicians have argued that it can be seen at the outset that such a search is futile. They have pointed to Gödel's famous principle of incompleteness and the related ideas of Turing and others to show that a complete and comprehensive reduction is not possible, even in theory.

Their point is perhaps most easily understood in terms of numbers theory. We are able to describe many real numbers with numerous digits to the right of the decimal point by a simple reductive formula. This formula allows us to describe the number without having to write it out in extenso, with all of its digits, to hundreds or thousands of decimal places or more. However, there are also numbers for which this is not possible. The only way to fully describe such numbers,

The Meanings of Reductionism

to fully entail them, is to write them out symbol by symbol, however long they may be. No reductive formula is possible, and as a result they are said not to be "compressible." This line of thought is not limited to numbers, but argues quite generally that we cannot expect to reduce phenomena in nature completely. There are always aspects that cannot be fully accounted for in a reductive algorithm. After Gödel, this is called the *largest possible theory*. From this perspective, the most encompassing theory, a theory at the limit, a theory of everything, cannot be found reductively.

Sometimes the rejection of grand universalism is thought to be identical to the rejection of the scientific goals of seeking generalizations and finding universal laws, and hence may be viewed as being antiscientific. But it is no such thing. It is one thing to believe that there are universal laws and that science should look for such laws through a process that seeks the greatest possible generalization, but it is quite another to expect to accomplish reduction to an all-encompassing algorithm.

On the other hand, it can be argued that the failure of grand universalism would falsify the whole reductionist enterprise. If grand universalism is the logical extension of the concepts of generalization and universalism—that is, if we seek to generalize as broadly as possible and if there are universal laws—then does it not follow that it should be possible to carry such generalizations to completion, in a single comprehensive law that encompasses everything? And if this view is incorrect, as some believe, then so is reductionism, because the reductionist edifice is built on the stones of grand universalism. If we are capable of making this or that generalization about nature, then, at least in a reductionist context, how would we go about placing limits on the extent and type of generalizations possible?

Lessons from the Living Cell

Dynamic Universalism

If reduction leads to fundamental laws, two significant questions arise: What do we mean by *fundamental*, and are such laws eternal? I will consider the question of the meaning of the term *fundamental* later in the section on parsimony as reductionism. The second question, whether such laws are eternal, can be thought of in two ways. First, are *nature's* laws eternal verities? That is, do laws of nature exist that are unchanging? The belief that they do is axiomatic to science and is implied by our use of the term *natural law*. Recent notions of multiple universes suggest the possibility that, like Newtonian physics, our laws may only be local. Still, the reductionist would argue, even universe-specific laws would share a common underlying solution that would transcend the particular circumstances of a given universe, much as general relativity encompasses Newtonian physics.

More relevant in the current context is the other reading of the question of whether fundamental laws are eternal. In this case, we ask whether *human-discovered* laws of nature are eternal. The dynamic universalist holds that although the natural world itself may have eternal laws, humankind's understanding of these laws is not fixed. It is constantly changing; hence, dynamic. The dynamic universalist holds that our current understanding of nature is just that—our *current* understanding. For example, our understanding of the nature of the physical world has changed greatly since Newton's time. We now live in a relativistic and quantum-mechanical world, in which questions of the relativity of time and space, of symmetry and broken symmetry, of chance and a predictable future, occupy the energy of some of the world's most thoughtful scientists. Change does not abate, reaching some final plateau of perfect knowledge. No understanding can be established as being true forever. Indeed, the dynamic universalist argues that if an idea were to be permanently established, it would become unscientific, even if we could not imagine how it could be false.

According to the dynamic universalist, it is central to science and its process that what we view as fundamental today may be incorrect, an anachronism, or a special case tomorrow, and this uncertainty applies to everything that we call scientific. One could fill a book with examples to show how human perception of what is fundamental has changed and changed again, year after year, century upon century. If we see relativity and quantum phenomena as more fundamental concepts than Newton's laws of motion today, will *they* still be as fundamental tomorrow? According to the dynamic universalist, who can say? In this view, to accept universalism is not to accept any particular assessment, past or present, of what is understood as being true or what is understood as fundamental.

Static Universalism

Some who argue the universalist perspective argue for a particularist and unchanging conception of fundamental laws. In this view, human understanding is seen as an accurate map of nature's own laws. The static universalist looks at today's knowledge as having been established as being true forever. This perspective is well put by a science writer as follows:

> *If one believes in science,* one must accept the possibility—even the probability—that the great era of scientific discovery is over. By *science* I mean not applied science, but science at its purest and grandest, the primordial human quest to understand the uni-

verse and our place in it. Further research may yield no more great revelations or revolutions, but only incremental, diminishing returns.[2]

The author seems to be arguing that new scientific theories, such as relativity or quantum mechanics, have only had value in the past. In the present era, when almost everything is known, progress can come about only by more modest means, by completing the almost completed crossword puzzle. There are no more general theories to be discovered, and no currently held views, at least no catholic views, to be overturned. To the static universalist, history serves as no exemplar for the present. The misperceptions of the past have no counterpart today, when our understanding has been, if not perfected, almost perfected.

Strong Static Universalism

Some thinkers hold to a particularly strong form of static universalism. They point to the remarkable success of science and ask, "Who is to say that there is any limit to what human beings are capable of uncovering, of understanding?" This seems a very optimistic view of human capabilities—optimistic, that is, if one believes that humans knowing everything would be an unalloyed good.

I personally find it hard to think of humankind's capabilities in such grandiose and egotistical terms. Modern evolutionary theory, among our accepted truths, does not present humans as standing godlike on the crown of a finished evolutionary tree, but as one temporarily successful sprig in the tree's multifaceted bushlike structure. Whatever humans'

---

[2] John Horgan, *The End of Science* (paperback ed.), Broadway Books, New York, 1996, p. 6.

intellectual capabilities may be, and they are admittedly prodigious when compared to those of other life forms on this planet, the process of natural selection seems ill suited to making them unlimited.

Reductionism as Atheism

For some, to reject reductionism is to cleave to the view that it is only through religious belief that we can account for nature. From this perspective, reductionism (science) and religion are incompatible opposites. In the rationalist world of the scientist, where faith is often given no quarter, equating antireductionism with the affirmation of religious belief may be no less than name calling. Whether or not this is true, religious affirmation is not what is in the minds of most scientists who question reductionism today.

Reductionism as Parsimony

A central tenet of science is that simpler is better. This idea was first articulated by Sir William of Ockham in his famous razor of parsimony. Sir William's razor states that the simplest hypothesis is to be favored over the more complex. Or, more accurately, it states that the theory that explains phenomena using the fewest principles is the most desirable, the most convincing, and the most beautiful. In a modern context, we say that a theory is fundamental if it provides the broadest possible explanation using the fewest parameters. For example, in seeking a unified theory for matter and energy, physicists seek to explain all of the known physical forces in the universe by reducing the number of parameters needed. This means explaining these forces in the simplest possible fashion.

This idea is related to reductionist principles. The atom is both simpler and more fundamental than the molecule in which it is found, and the molecule is simpler and more fundamental than the animal or plant in which it exists, and so on. Thus, the simplicity postulate in a reductionist context argues that simpler is deeper or more fundamental. Thus, in the application of the fewest parameters to provide the broadest possible explanation, we are drawn to the conclusion that complex systems may be interesting and indeed very important in the lives and times of our species, or even of all life and all nature, but they cannot be fundamental. In this view, the search for fundamental laws of complexity is the search for an oxymoron.

But there are problems with such a conclusion. Certainly, an algorithm that really explained everything would have to include everything—that is, the informational, contextual, and relational aspects of things, as well as their physical embodiments. Accordingly, such an algorithm could not be simple. Indeed, it would seem necessary that it be of great, perhaps even unimaginable, complexity. Thus, the only way to develop an algorithm that is both expansively general and yet simple would be to exclude any and all complicating aspects of reality from our description. Even though we might argue that these are trivial and inconsequential factors, nuisances that can be ignored, can a theory really be complete without accounting for them?

In any event, most scientists, whatever their views of reductionism, hold to the notion of parsimony in theory. I say *in theory* because, at least in biology, in practice Ockham's razor is more often ignored than honored. It turns out that in biology complex explanatory theories are often proposed at the outset. The vesicle model is an example of such a theory, and we will discuss the how and why of it later on.

When such theories are confronted with difficulties, they are not likely to be rejected, and if popular enough may not even be modified. And when modifications are made, the theory invariably becomes *more,* not less, complex. At times there seems to be no limit to the acceptable accumulation of complexities, if the alternative is rejection of a favored theory.

Moreover, unlike mathematics, in biology complex explanations may be perceived as beautiful and elegant, whereas simple ones may be rejected as naïve and simplistic. The reason for this is that living systems are the most complex known to us, and we are not surprised when we encounter complexity in them. And it is at least in part in these complexities of life that our ideas of beauty are grounded. Thus, in studying biological systems, we are not only not surprised when we find complexity, but often see beauty and truth in it.

Reductionism as Objectivity

Reductionism is often equated with an objective description of reality. Accordingly, human perceptions that do not conform to a specific reductionist view may be seen as imaginary, or not objectively real. If one experiences something—a thought, a feeling, an understanding—that cannot be explained in reductionist (neurobiological) terms, then such experiences are illusory. Or, if they are real, it is nonetheless as if they were imagined, because we cannot "objectively" confirm them.

A central outcome of Descartes's separation of body and soul was that we could understand nature "scientifically" only by what we usually refer to as *third-person observation*. This is called *objective observation,* to contrast it with subjective or first-person reports. Traditionally, science has accepted only examination by parties that are external to the events being examined, and as such are considered independent of the phenomenon itself. Our confidence in scientific observation is in great measure a consequence of this understanding.

Subjective perception cannot be achieved independently of the phenomenon being examined. Both the phenomenon and its report derive from the same source. This, it is argued, means that we cannot dissociate the narrator's personal prejudice or potential misjudgment from his or her report of the phenomenon. And as a consequence, we cannot have confidence in the validity of such reports.

But, of course, observers who are objective in the sense given may display prejudice and misjudge what they observe, whereas a reporter of subjective experience may be scrupulously honest and display extremely good judgment. The difference between the two lies in the fact that for subjective observation we must rely on the word of the reporter, the person from whom the phenomenon arises, because no other source is possible; whereas, when the observer is external to the phenomenon, one person's report can be compared to that of another.

We call this characteristic of objective observation *reproducibility*. It has two aspects: the ability of additional examiners to report the same phenomenon independently, and the ability of the initial reporter to repeat his or her findings. Although science relies heavily on both, reproducibility by others is defining. All sorts of remarkable things are reported that can never be validated by others and, as a consequence, remain outside scientific understanding. And yet, in spite of this need to reproduce phenomena to evaluate them scientifically, certain singular, historical events are central to understanding nature, such as the origin of the universe or of life, and remain part of science in spite of their uniqueness and apparent irreproducibility.

In any event, objective observation is not the sole provenance of reductionism, unless by reductionism we simply mean science itself. One may reject any reductionist proposition and still carry out objective observations. Reductionism has no special claim to either objective observation or objectivity.

Reductionism as Materialism

Materialism says that everything in the universe is matter or void, something or nothing. Because reductionism has concerned itself in great part with learning about matter, materialism is sometimes considered an aspect of reductionism. But one can be a materialist— that is, one can believe in the existence of a material world—without necessarily holding reductionist beliefs. Nor does holding such beliefs exclude one from understanding that there are aspects of the universe, such as space, time, and information, that are not material in themselves, and yet are not without meaning and importance. Beyond this, as we have probed deeper and deeper into the nature of matter, its material aspects have become more and more evanescent and uncertain, and what we mean by *matter*, as opposed to *energy*, has become more elusive.

Reductionism as Determinism

Determinism says that if we understand a system well enough, then we can predict with perfect accuracy its future behavior. That is, if we can specify the initial conditions, then the effect of a particular cause can be fully understood or determined.

Even though determinism, in the main, still holds sway in physics, where it has been most fully realized and where it has dominated thinking for centuries, it has nonetheless been under fire for most of the twentieth century. Quantum mechanics has shown that there is apparently much that is unpredictable even about atoms. For example, we can only predict the location of an electron in orbit statistically. At the quantum level, chance, not certain prediction, leads to outcomes. As a consequence, the location in space and time of a fundamental particle cannot be determined in a predictive fashion, but only by measurement.

Being able to accurately predict the future is perhaps the most remarkable feature of Newton's laws of motion. If one thinks about it for a moment, how remarkable and counterintuitive this is becomes clear. We cannot predict the future with much accuracy in everyday life, though we try. Our lives are made of various uncertainties, large and small. Whom we will meet, what decisions will be made, how long it will take us to get to work, when we will die and of what, and on and on and on. Yet we can predict with unerring accuracy the location of a spacecraft in a minute, an hour, or a month.

In this light we view the behavior of the object flying in space as being more fundamental in nature than the course of our own uncertain lives. Still, to the determinist the uncertainties of life are only apparent. They only seem uncertain because we lack the necessary information to completely understand the complex array of causes. If we could list all the causes, if we knew the initial conditions fully, then we could show that our future is indeed determined, like that of a thrown rock. Then we could even put Nostradamus to shame.

The Meanings of Reductionism

The determinist argues that for biological systems, input from the environment wholly determines output from the organism. Seeing someone you recognize and moving toward him or her, or seeing a precipice and avoiding it, or engaging in the most complex human mental activities you can imagine, are all simply the result of environmental input, through the intermediate step of integration by the organism, determining the inevitable outcome in a fully causal manner. No intervening conscious analysis and decision making really (objectively) occurs. Notions of free will and conscious choice, even of consciousness itself, are mere illusions.

Those who hold this view often see themselves as reductionists, and consider those who believe in the "I" of conscious thought and in decision making by humans and animals as being devoted to incorporeal substances and magical thinking. But believing in the existence of conscious mental states or the intentionality of individual action does not make one nonscientific or even antireductionist. There is no contradiction between the view that there are conscious mental states and that we make intentional decisions, and the view that these activities are based solely in the universal reductionist properties of neurons.

Weak Microreductionism

With Democritus, microreductionism proposes that all material objects are composed of smaller underlying elements, and that by studying these smaller elements we obtain the most fundamental or deepest knowledge. Thus, microreductionism relates the idea of fundamental knowledge to the size of the material being examined. By probing smaller and smaller entities, we obtain the most fundamental knowledge.

As stated previously, in this sense the molecule is more fundamental than the object in which it is found, such as a biological cell. The atom is more fundamental than the molecule, the subatomic particle is more fundamental than the atom, and so on to an uncertain end. Smaller is always more fundamental than larger. From this perspective, the highest goal of science is to probe ever more deeply into the substructure of matter. Moreover, microreductionism is related to the search for simplicity, because smaller is simpler as well as more fundamental.

That underlying aspects of matter exist that are more fundamental in the sense meant and simpler than the larger structures from which they are derived is well established today. They do and they are. But for the life sciences this ancient and widely validated notion of microreductionism has become something quite different and far more questionable. It is when we consider microreduction in the world of the life sciences that we come upon the important difficulties with reductionism that are the central concern of this book.

Strong Microreductionism

In biology the weak form of microreductionism has more and more commonly been replaced by a strong version of this proposition, which argues that we can come to understand all phenomena completely from knowledge of their underlying structures, their constituent parts. Indeed, in this view, this is the only way we can obtain a rigorous understanding. This means that everything about the larger object can be attributed to its parts. Or, put another way, the object, whole and intact, is fully entailed, or caused, in all of its aspects by its substituent parts.

This may seem trivially correct if we restrict ourselves to structure. A whole structure, whatever it may be, is the sum of its constituent parts. Could it be anything else? But once established on this footing, strong microreductionism claims much more. It asserts that all conceivable properties of the whole—for example, its relational, contextual, and informational features—are attributable to or entailed by the properties of the component parts. It argues that *the whole has no properties beyond those of its component parts.* Taken to this extreme, it maintains that the fact that we cannot describe humans, no less human society, in terms of their constituent atoms is simply a technical difficulty. There is nothing fundamental that stands in our way. However inconceivable this may seem, if we could specify the atoms completely, such a description would be provided.

Early in this century the mathematician David Hilbert began an enormously ambitious research program. He argued that if reductionism is correct, then it should be possible to explicitly formalize all of mathematics in a series of expressions. If this could be accomplished, then an algorithm could be developed to produce mechanical answers to any and all aspects of mathematics. Hilbert failed, and the death knell of his research program was tolled by Gödel's incompleteness discovery. As we have already discussed, even in the simplest areas of mathematics, such as numbers theory, a reduction is not always possible. Gödel's theorem was called the *incompleteness theorem* because whatever the reduction, and however well one could entail a particular theory, something was inevitably missing, and that something was necessary for the most encompassing, the most general, explanatory model. One implication of this conclusion is that there is something beyond the parts of systems that can never be specified, can never be entailed from knowledge of the parts alone. As such, a fully descriptive reductionist algorithm just cannot be written.

As we shall see, what is missing from this reductionist view are the properties of objects whole and intact. In biology there are phenomena that originate only from whole cells and organisms. They involve the relationships between various parts of the object, the context in which it exists, and any information that it might contain, all expressed in space and over time. These phenomena cannot be found in, nor understood or inferred from, the intimate structure of the object—that is, from the properties of its various parts viewed in isolation—however rigorous our knowledge of them may be.

The Strong-Microreductionist Research Program in Biology

As already suggested, strong microreduction is not only an important philosophical concept, but it has become the dominant methodological paradigm for research in the life sciences. As such, it demands allegiance even from those researchers who do not consciously subscribe to it. Biologists applying *weak* microreductionism attempt to break open the various black boxes that compose living systems in order to isolate and identify their underlying structures and examine their most deep-seated properties. They argue that such knowledge is *prerequisite* for a rigorous understanding of the whole. It is a necessary but not a sufficient condition.

In practice, however, it is difficult to know where the boundary lies between weak and strong microreductionism. If a biologist acquires thorough knowledge of the particular molecules that compose a subcellular structure, he or she will then be compelled to provide a complete description of it. Or, moving up the scale of complexity, if a biologist is able to elucidate all of the major substructures in a cell, then he or she should be able to provide a complete description of that cell. And complete knowledge of the cells that compose organs and tissues is accepted as a comprehensive description of that organ or tissue. From this the belief seamlessly follows that the organism itself can be understood in these terms, as the sum of its constituent organs and tissues. From the perspective of the strong microreductionist, self-professed or unwitting, the parts entail the whole in all of its aspects, whether we are talking of cellular substructure, cells, organs, organisms, or even groups of organisms. And if this is true, we arrive at a remarkable conclusion. If each element in this chain of life is the sum of its component parts, then life can be reduced to its molecular instantiation. That is, life not only has a molecular basis, but complete molecular causation.

Ultimately, from this viewpoint an essential, if not completely comprehensive, description of life does not even require the enumeration of all of the included molecular parts. As already discussed, life can be reduced to two molecules and two simple truths about them—protein molecules entail biological function, and DNA molecules entail proteins. In biology there are no "incompressible" phenomena, and the living state can be explained entirely at the molecular level by this central "algorithm" of life.

I have used the word *entails* to mean "describes or accounts for fully." Because its meaning will play an important role in what follows, it is useful to take a moment to be more explicit. It is a very important word. Most simply, it means that A causes B for objects, or A implies B for models. Beyond this, it means that A is necessary and sufficient for B. Nothing else is required for B. In a strong-microreductionist context, this means that the whole (B) is the sum of its parts (the collective A), and nothing more.

The Use of Strong Microreductionism

Though it is widely applied, the strong-microreductionist viewpoint is usually left unstated or is not stated directly. Occasionally, its echo can be heard in criticism that particular research lacks rigor or stringency because it lacks a strong-microreductionist approach—that is, because it does not attempt to achieve understanding of the whole by studying the chemistry or physics of its substituent parts. Or in an overarching confidence that the whole either has been or will inevitably be illuminated from such knowledge.

Strong microreductionism has been so influential that it is often equated with reductionism per se; consequently, biology's great successes are sometimes attributed to it. And *a circulare,* such attributions serve to validate the proposition's view of the nature of life and how to learn about it. But when we examine the accomplishments of biology, paying attention to the full palette of knowledge that led to them, we see that *none* is a consequence of strong microreductionism.

CHAPTER 5

Epistemon and Eudoxus

Humpty Dumpty sat on a wall,
Humpty Dumpty had a great fall;
All the King's horses, and all the King's men
Cannot put Humpty Dumpty together again.

Mother Goose

Mother Goose's simple nursery rhyme presents us with a profound conundrum. Why couldn't all the king's horses and all the king's men, with all their combined power, their financial and technological resources, put this simple egg—white, yolk, and shell—together again? Furthermore, why couldn't we do it today with all of modern science and technology at our disposal? The Humpty Dumpty conundrum is emblematic for biological beings and fundamental to understanding the nature of life. Much of what I have to say in what follows is an attempt to answer this question about poor Humpty's fate.

Lessons from the Living Cell
Life as a Self-Organizing System

The modern reduction of life to the agency of DNA takes us beyond cell theory, much in the same way as subatomic particles such as bosons and quarks take us beyond the atom. It is clear today that DNA replication is the central event in the production of new life. And, as said, whether we talk of their chemical, physical, or structural properties, proteins are the central actors in the life of the cell, and the life of organisms more generally. It is unmistakable that life does not exist without them. And while proteins are necessary for life, DNA is necessary for proteins. This has all been established.

From here it is a small step to the claim that DNA is not only necessary for life, but sufficient for it. That is to say, through its actions and functions, and by way of them, through the actions and functions of proteins, life is brought into being. This conclusion rests on two fundamental understandings. First, DNA provides all of the information necessary to produce life. And second, this information is limited to specifications for the structure and expression of the cell's protein molecules. In other words, DNA entails proteins, and proteins entail life.

This is the ultimate or the extreme, depending upon one's point of view, conclusion of the reductionist agenda in biology. In this estimation, the other material elements of life—other molecules, cellular substructure, tissues, organs, even the organism itself—arise as part and parcel of the self-entailing system that begins with DNA. This is life as a self-organizing entity in which the parts provide all of the information necessary to account for the whole object. As such: if the cell is entailed, fully accounted for, by its molecules, if cells produce tissues and organs, and if tissues and organs compose the organism, then we (as organisms) are simply the additive product of our component parts. From this perspective, the material parts of cells are not only necessary for life, but in and of themselves produce it. This understanding leads to the following categorical prediction of the strong-microreductionist principle. If the parts account for the whole, and nothing else is needed—then under the right circumstances, that whole would be expected to organize itself spontaneously from its parts.

Can the Cell Assemble Itself?

In an unfinished dialogue, "The Search after Truth," René Descartes introduces us to two scholars, Eudoxus and Epistemon, "the most distinguished and interesting of their time," the first of whom "has not studied at all," while the second "is well acquainted with all that can be learnt in the schools." Eudoxus characterizes Epistemon to their host Polyander in the following way, and in so doing, characterizes himself as well:

> I dare not hope that Epistemon will give way to my reasoning. He who is like him, full of opinions and prepossessed with a hundred prejudices, finds it difficult to hand himself over to the light of nature alone; for long he has been accustomed to yield to authority rather than to lend his ear to the dictates of his own reason.
>
> In what follows I present a dialogue between two modern scholars much like Epistemon and Eudoxus, in which they contemplate whether under the right conditions the cell would assemble spontaneously from its component parts. I have taken the liberty of calling our modern scholars Epistemon and Eudoxus after Descartes. Our Eudoxus is a most

talented experimentalist and a recent convert to biological research from the physical and chemical sciences. Although he is a thoughtful and well-trained scientist, he lacks much detailed knowledge of biology. Epistemon, on the other hand, is a respected experimental biologist with many years of research experience who knows all worth knowing in his field. In addition, he is a stalwart strong microreductionist. Although our scholars' names hark back to Descartes, they are in every other way the very models of early twenty-first-century scientists.

Eudoxus and Epistemon occupy laboratories down the hall from each other on the same floor of a modern research building at one of our most respected institutions of higher learning. Their dialogue began as the result of an experiment performed by the biologically untutored Eudoxus. To say that his experiment was ambitious would be an understatement of major proportions. Eudoxus was trying to understand nothing less than the fundamental nature of cellular life. He had devised an experiment to test the conclusion that the molecular contents of cells fully entail life, and consequently that the biological cell is a self-assembling or self-organizing system.

His plan was simple, although carrying out the experiment was not. He would take a cell, break it open, collect all of its parts, and then separate them from each other, substance by substance, molecule by molecule. Having accomplished this enormous technical feat, he would then take the separated material and put it all back together in a special device that he had invented for this purpose. The device was special because it was able to mimic the environment in which the various substances had existed prior to their separation. It allowed Eudoxus to control temperature, pressure, volume, and all other imaginable variables of state. If the parts were all the entailment needed, then under these conditions the cell should reassemble spontaneously, or so Eudoxus hypothesized. If he was able to demonstrate self-assembly, he would go to the next step and try to form tissues and organs from their component cells. Some day, he imagined, he might even be able to assemble whole organisms.

Separating the cell's multitudinous elements and then putting them back together under the right conditions was certainly a difficult task. As best Eudoxus could tell from the literature, it had never been attempted. Nonetheless, after much effort he was able to completely separate the parts of the cell. Then, with much attendant discomfort given the prodigious effort of separating them, he put them all back together again in his special device.

He assessed the success of his experiment by analyzing the contents of the fluid broth in his device at various times. Almost immediately, a variety of insoluble structures appeared. They were molecular aggregates—micelles of fat; precipitates of protein; or combinations of fat, protein, and all manner of other cellular substances. This gave Eudoxus confidence that he might succeed, and that more and more complex structures would emerge and sooner or later coalesce into whole cells. But time turned out to be unfriendly. Even though he continued to examine the contents of the broth, the hours turned into days, and the days into weeks. No matter how long he waited, the hoped-for structures never emerged, and certainly, no living cell.

Eudoxus was disappointed to say the very least. After all of this effort and all of his hope, his experiment had failed. But being new to biology, he realized that he might have missed something, some important detail. He decided to discuss his experiment with an experienced biomedical scientist. Even though he knew Epistemon only slightly, he realized that he was highly regarded by his colleagues and possessed a wealth of knowledge. Since they were neighbors, he decided to approach him.

Epistemon was shocked when Eudoxus told him about his experiment, no less his dream of building tissues and organisms from their component parts. How terribly naïve, he thought. After a few uncomfortable moments he responded.

EPISTEMON: You should not be surprised with your result. With the exception of certain viral particles, and to some degree bacteriophages, the reassembly of biological cells has never been achieved in the laboratory. Nor is it obvious that it is a reasonable expectation, or even a conceivable one. That is why you found no citations in the literature. Self-assembly is quite unlikely, if not impossible, even for the simplest cells. And to think that if you could somehow reassemble a cell, that this could be extended to the manufacture of whole tissues and organs, no less to complex organisms, all from their component parts—like Dr. Frankenstein or some *Star Trek* transporter—is patently absurd. It is pure science fiction. Such a goal is just not achievable. *(Pausing for a second)* But you should not take this judgment to mean that the parts, all of which you seem to have in hand, do not entail the whole. They most certainly do. There is nothing else.

This confused Eudoxus. With this last statement Epistemon seemed to have contradicted what he had said before.

EUDOXUS: If the parts fully entail the whole, then how come reassembly is not possible? Doesn't one follow from the other? If no other causes are involved, then shouldn't the cell simply reassemble, reorganize itself, from its constituent parts?

Epistemon and Eudoxus

EPISTEMON *(patiently)*: Reassembly is only possible in theory, not in practice. Eudoxus, your experiment was ingenious enough. Indeed, I am impressed with your skill. But I am afraid that however ingenious and skillful you are, your research project is doomed to failure. You will not get funding from the NIH for such a project—and I hope you will not think me too forward, but I would advise you to change your research focus. This will get you nowhere.

I see that I have perplexed you. So, let me be as clear as I can. The reason that you were unable to reassemble a cell in your special device was not because you were missing something *fundamental*. There is no entailment beyond the parts themselves that you could not or did not provide. None exists. The problem is that there is much that is not fundamental that stands in your way. To produce reassembly you must be able to specify—and specify *fully*—all of the initial conditions, and you are in no position to do that. They are manifold, and you are not even close to having the requisite knowledge. This is not your fault. The information is simply not available. If it were—a faint hope today even with all of our knowledge—then, yes, perhaps the cell would reassemble spontaneously.

But in spite of our inability to demonstrate reassembly in the laboratory, there is much that tells us that the cell is indeed nothing more than the sum of its component parts. You should not think otherwise. For example, we know that for simpler systems, such as uncomplicated viruses, where we can adequately establish proper initial conditions, reassembly can occur. In addition, supramolecular structures—for example, homologous [composed of the same type of molecules] and heterologous [composed of different types of molecules] aggregates of proteins, such as microtubules—are easily reassembled today. What needs to be done is to extend such observations to more and more complex entities, step by supramolecular step. Then, eventually, with much effort and patience, and no doubt after many years, perhaps even centuries, we may be able to reconstruct the whole cell. But first we must establish all of the necessary preconditions, and that is still far beyond us. Yet, however far we have to go, this limitation is merely technical, however intimidating it may be. I implore you not to mistakenly conclude that your failure is a consequence of some fundamental outstanding question or questions. It is not. No doubt when we have accumulated all of the necessary knowledge, the elegance of nature, embodied in the structure of its biomolecules, will be revealed as the cell in fact reassembles.

In the following days and weeks Eudoxus thought much about this conversation with Epistemon, particularly his insistence that the problem was technical. Eudoxus realized that he could not be certain that the conditions he had chosen were exactly right, and he thought that he could deal with many of the specific problems that Epistemon had raised. Being an obstinate soul, he decided to attempt new, far more extensive studies, rather than give up his interest in the question as he had been counseled. If the problem was technical, well, he would just have to overcome it.

Rather than choosing a particular set of state variables as he had done in his initial study, this time he would alter all of the variables independently of each other—temperature, pressure, volume, chemical contents, and so forth. He would then alter them in combination, for all possible combinations. In this way he would eventually find the proper initial conditions. Though the prospect of such an experiment was daunting, this seemed to be the logical path, and he took it. But it turned out that after an immense effort, after carrying out many, many experiments, he could not find any condition or set of conditions in which cells emerged from their component parts. He learned many new things, and found many new and interesting complexes, but still no cell.

By now resigned to his failure, Eudoxus approached Epistemon again, this time somewhat sheepishly. Epistemon was not as patient as he had been during their first interaction. He had given Eudoxus good advice, and it certainly was no surprise to him that Eudoxus had failed.

> EPISTEMON: However many states you have considered, and however many variations on those states you have tried, obviously—and, indeed, by definition—you have not found the right one. It is there to be found, but you cannot find it, because you do not have the necessary knowledge. And I am confident that you will never find it, however many experiments you do. I do not say this to be critical or argumentative. What other choice is there, really? Why else would you fail? What other reason could there be?

With this question Epistemon had joined the issue. He was saying that since we have in great part described the material contents of living systems—certainly those that are central to life, all of which Eudoxus had in his brew—what potential source of entailment could be missing? An immaterial substance, a metaphysical force?

EPISTEMON: That the molecular contents of cells fully entail the living state is certainly the simplest explanation, and it appears to be the only one. There is no reason to suspect anything else, and much reason to suspect that there is nothing else. Even if you could offer another explanation, it would necessarily be more complex, probably far more complex. Science, as you know, must hew to the path of simplicity in the absence of convincing evidence to the contrary. If I am wrong, and the problem is not just technical, then offer an alternative!

The Skeptic's Response

By now Eudoxus realized that he had been naïve about the possibility that his experiment would succeed. But neither his lack of success nor Epistemon's certainty about its cause had convinced him that there was an unknown and unspecified technical explanation for his failure. He had become a skeptic. He believed he had achieved a meaningful negative result and thus could no longer accept the premise that had given rise to his study in the first place. He now found himself asking whether the cell is really a self-organizing system.

His conversations with Epistemon continued from time to time, but their tone and character changed. Things were reversed. It was now Eudoxus who was asking Epistemon to examine his position. It was Eudoxus who was asking probing questions. He pointed out that he did not have to know the nature of the missing entailment or entailments before he could question the notion of self-assembly. He simply had to show that something was missing, whatever it was. Nothing more. It seemed to him that it was simply a fact, a fact that neither he nor Epistemon could ignore, that even with the most abundant information and his careful and comprehensive experimentation, he had been unable to reconstitute a cell, reconstitute life. Not only that, but Epistemon agreed that his failure was assured. And although he insisted that this was due to technical limitations, he could not specify what those limitations were.

EUDOXUS: How do you know you are right? Aren't you simply *presuming* that your explanation is correct? Isn't it in fact unsubstantiated, however parsimonious and reasonable it might seem to you? Epistemon, your confidence seems to be based on little more than your indefensible trust in an unproven hypothesis.

And certainly whatever may or may not be missing, my experimental evidence can never be taken to support your view. Nor have any others reported cells, much less a more complex living thing, being reconstituted spontaneously outside of fiction, as you have pointed out. You say that we can reassemble simple viruses and subcellular complexes of various sorts, but cannot expect to reassemble a cell. The question is *why*?

I am mindful that I may have difficulty publishing my negative results, because the reviewers will seek some technical explanation for my failure, however uncertain and unspecified. They will say that the experiments are "technically flawed." But such a conclusion would simply be ad hoc. And strongly held views should not be confused with valid ones.

Immaterial Entailments

EUDOXUS: You argue that only a technical explanation is possible because the parts *must* provide whatever entailment there is, inasmuch as there is no other rational possibility. It was in this light that you challenged me to offer an alternative. I do not admit to it being a fair challenge, since the weakness of your position stands or falls on its own. If I have no alternative to offer, it merely points to my ignorance, not the strength of your position. Nonetheless, I accept the challenge.

You say that there is no option. But you are wrong, and obviously wrong. The physical world cannot be fully explained in terms of its material contents. As I am sure you realize, two central aspects of physical reality have no material embodiment whatsoever: space and time. I suggest to you that my inability to reassemble the fractionated cell was due to my inability to specify and effectuate time and space constraints. They are the missing entailments, and they are not to be found in the molecules themselves. The difficulty after all, dear Epistemon, is fundamental, not technical.

Space and Time

EUDOXUS: By *space*, I of course refer to the three-dimensional space filled by the cell and organism. Even in a simple cell, its components are "compartmentalized." That is, they are not randomly distributed throughout the object, but found at distinct loci. I am concerned primarily with proteins, though the same principle applies to other substances. Even the simplest bacterium contains two compartments: the cytoplasm and the cell membrane. The cell's proteins are fated to either location, or they may be released from the cell into a third compartment, the environment. That is to say, even in these simplest of living things, proteins are ordered in regard to their location. They may be in the cytoplasm or membrane, or be secreted substances. Many bacteria contain an additional (outer) membrane, and hence a space between the two membranes, an intermembrane space. In these cells proteins are "sorted" or *targeted*, to use the language of the field, to five different sites.

As you certainly understand, the situation is far more complex in the cells of higher plants and animals. Here the number of cellular compartments is greatly increased, and these compartments often have subcompartments. For example, a protein destined for the nucleus may be located in its membrane or its internal matrix or be associated with any number of specified nuclear elements, such as DNA, RNA, or other proteins. In polarized cells, proteins in the membrane may be sited at one end or the other of the cell, or at different specific loci along its perimeter. Mitochondrial proteins are not simply sent to the mitochondrion, but may end up in its outer or inner membrane or its matrices. Even many cytoplasmic proteins that were once thought to be wholly unordered, floating around in a solvent soup, are now known to exist in ordered arrays at particular locations. For example, complex webs of linear protein polymers, called *microfilaments* or *microtubules*, crisscross the cytoplasm and attach to membranes at specific sites.

Although we cannot stipulate the exact coordinate locations for each and every protein molecule in the cell, every species of protein has a home, and often more than one home. We can think of these locations as fuzzy sets—such as membrane, mitochondrion, cytoplasm, or nucleus—that have topologic meaning even while lacking exact coordinates. This specificity is required. Without this spatial entailment, there can be no living cell.

EPISTEMON: I agree that entailment for location is necessary for life. But you have certainly chosen a bad example to make your point. We know today that the information necessary to site proteins at particular locations is found in the structure of the protein molecules themselves. It is another example of nature's simple beauty. It has been shown, at least for some proteins, that specific sequences of amino acids target them to particular locations in the cell—to the nucleus, rather than the mitochondrion; to the nuclear membrane and not its matrix; to the cytoplasm and not the cell membrane; and so forth. Whether these sequences provide a code or signal for location, as is commonly thought, or represent more general structural features of the proteins, it is *their* structure and their structure alone that accounts for their eventual location. No external referents are necessary. No additional entailment is needed.

EUDOXUS: But you have not really provided a *cure* for the problem. In my experiment, the protein, however topologically well informed, had no home to seek. There was no mitochondrion, no nucleus, not even an inside or outside of the cell to go to. It is not a matter of simply finding the right place in an already ordered environment, but of finding the right place in the absence of prior order. My experiment began *without* order. That was the point. If you are correct about self-assembly, then the molecules must be able to create that order, that organization all by themselves. This is quite different from a whole cell with its already established organization.

To produce complex cells from scratch that contain a cell membrane, ribosomes, nucleus, rough and smooth endoplasmic reticula, lysosomes, peroxisomes, secretion granules, and mitochondria, to name a few important elements, and all in the right locations, presents topologic complexities that are beyond imagining. They certainly cannot be resolved by molecular sorting signals. There must be other entailment.

But however problematic the question of spatial location may be, it is the question of time that seems truly insurmountable to me. From what we know of interactions and reactions between biomolecules, that any random sequence of events could or would accomplish the assembly of a cell seems beyond reason. For assembly, interactions must occur in a specified, or at least more or less specified, sequence. And this knowledge is not to be found in the protein molecules themselves. Certainly, one can imagine how interactions between two molecules might enhance their tendency to associate with a third, and so on. But it seems ludicrous to think that an extended and branched chain of such events, none of which are all or none, yes or no, would emerge spontaneously from among the various molecules in our broth and produce a whole cell. In the *formed* cell, the timing of protein expression is regulated in various ways by means of specific regulatory segments of DNA, but this is irrelevant in my experiment. All of the proteins are already present in the mixture. New expression is immaterial.

It seems to me, Epistemon, that unless you can provide me with a reasonable theory—or, better yet, evidence—to support the hypothesis that the sequence of events needed to form a living cell, properly sorting all of its chemical components, is wholly entailed by its molecular structure, then a living cell must be more than the sum of its parts. Additional entailment, additional causation, is required. This tells me that your strong microreductionist view is incorrect. Neither cells nor organisms are self-organizing entities in this sense. They are not wholly entailed by their material substance. Supplementary entailment is needed to deal with the problems of space and time.

The Cell, the Machine

EUDOXUS *(continuing):* Occasionally, on television or in the movies, a large building is shown being demolished by explosives. Sometimes the footage is run backwards, and the viewer can see the building rise up from the rubble by itself. This resurrection is supposed to be humorous or amazing, because we know full well it is not possible. Similarly, if we were somehow to find all of the separated parts of an automobile on the floor, close to each other, even touching, and in proper juxtaposition in all three dimensions, we would have no expectation that they would ever assemble themselves.

Epistemon and Eudoxus

True to your strong microreductionist beliefs, with its Cartesian proscriptions, you view biological systems as machines. And yet you expect that in animate—that is, living—machines, as opposed to inanimate ones, the parts in themselves entail all. In the same way that it is intuitively obvious that automobiles and buildings do not self-assemble, shouldn't it be equally clear that cells and organisms cannot, if the machine analogy is apt?

Automobiles and everything else that we call machines, other than living machines, are products of human intelligence. They are human constructions and as a consequence can only be assembled by humans or other human constructions. We provide the additional entailment necessary for their assembly. But if humans provide that entailment for machines, who or what provides it for living things?

EPISTEMON: Your point is well taken. Perhaps the simple machine analogy fails. Still, experience tells us that many things in nature, even inanimate objects, can "assemble" spontaneously. Perhaps such complexes are more like living objects than Descartes's mechanical machines. For example, a suspension of molecular aggregates insoluble in water at a given temperature may disassemble (become soluble) when the temperature is increased. The aggregates can be reassembled (precipitated) simply by lowering the temperature. And if the temperature fluctuates between hot and cold, the structure will continuously come in and out of solution, forming and disassembling time and again. Or take the solubility of salts in water. We can "reassemble," even crystallize, a solid salt simply by evaporating the water in which it is dissolved. And of course it also can be solubilized and extracted from solution time and again. Multitudes of substances assemble spontaneously, given the proper conditions.

EUDOXUS: It is certainly true that the formation of precipitates and like natural structures in the inanimate realm are entailed by their component molecules in a way that is not the case for human constructions such as buildings and cars. Nonetheless, even though your example is well chosen and such natural structures have important similarities to living things, they are far, far simpler. Indeed, *the reason for their simplicity is that they lack the entailment we seek for living systems.* And do I understand you correctly, Epistemon? Are you saying that we are not machines?

EPISTEMON: Yes, Descartes was wrong about humans and machines. Nature does not produce mechanical machines. They are only human-made, as you say. But surely you are aware that, unlike your building, reassembly does occur for living things. We have already talked of self-assembly at the supramolecular level—isolated molecules, most significantly various proteins, come together spontaneously under the right conditions. Protein molecules associate with each other in vitro, forming homologous and heterologous aggregates of various sorts that appear to be both structurally and functionally comparable to structures that are formed from the same molecules in the cell.

Epistemon and Eudoxus

Indeed, you found that such structures emerged in your experiments. Even though you failed to reach your ultimate goal, if some isolated parts of cells can assemble spontaneously, why shouldn't we expect that with additional effort to get things right we will be able to produce more and more complex entities in the same fashion? Perhaps eventually we will be able to form relatively elaborate subcellular structures this way. Even the cell itself is not beyond our reach. Although, as I have said, this is obviously far in the future, I see no theoretical barriers, simply difficult practical ones. I do not understand what it is that you think bars us from eventually realizing such a research program, other than the intimidating complexity of the undertaking, to which I admit freely.

If we had the necessary knowledge, we could indeed reconstruct your cell in the laboratory, and *without any additional entailment*. First, having established the proper sequence of events, we would design our system accordingly. If instead of the simple-minded single-beaker experiment you carried out, we used a series of beakers, each with its own specified state, each containing different substances, and all of them to be mixed and sorted in some specified sequence, we would stand a far better chance of success, or at least of progress. If we knew how to carry out this mixing and sorting, then the cell would indeed arise spontaneously. Someday it will be done, and a Nobel prize will be awarded for one more success of the strong-microreductionist enterprise.

EUDOXUS: Nobel prize or no Nobel prize, new strong-microreductionist achievement or not, you ignore your own hand, the human hand, in the enterprise. We could not conclude from the success of such an experiment, as remarkable an achievement as it would be, that the problem was simply technical. Such a cell would not have assembled itself. Its assembly would be the product of human intelligence. It would have been assembled by the investigator, just like a mechanical machine. If the molecules embodied all of the necessary information as you argue, then we would not need a complex system of mixtures and transfers for reassembly to occur. The fact of the matter is that proteins, DNA, and all of the other organic contents of cells necessary for life are not sufficient to produce it in and of themselves. An external hand is required, and in your experiment, it is a conscious and informed human hand.

EPISTEMON: But you must admit that if and when this experiment is carried out, we will have manufactured a living cell in a fashion wholly analogous to the way we manufacture automobiles. Human hand or not, the cell is a machine, albeit a complex and sophisticated chemical machine. There is nothing else. Moreover, if you think about it for a moment, you will realize that self-assembly *does* occur in whole living systems. Organisms display this capability in and of themselves. For example, you get a cut, and with time the skin heals, replacing what had been lost or damaged *by itself*. A salamander may lose a leg, but not to worry, *it* simply replaces it. Or finally, in exciting new gene therapy studies, a gene for a factor that causes capillary proliferation is injected into heart muscle, and lo and behold, new capillaries are formed. Nothing else is needed, just the gene, just the DNA.

Epistemon and Eudoxus

The Blueprint and Transgenerational Entailment

EUDOXUS: Epistemon, you have been blind to the obvious. The evidence, or more properly the lack of it, requires that you accept the conclusion that the parts do not entail the whole. Other forces are required.

If cells do not self-assemble, and sad experience tells me that they cannot, then they must be assembled by others. In nature, if not in your multibeaker experiment, this assembly is not due to a human hand, but some other cause. What is needed is not unlike what we need to put a car or building together. For a car or building, we must have the necessary parts, human hands (or hands made by human hands) to do the actual work, and the human mind in the form of a blueprint that specifies what to do with the parts—the where and when of assembly. Where is the blueprint for the cell, for the organism?

The answer is not only apparent, but central to cell theory. If buildings and cars require the intervention (and invention) of the human to come into being, then in an analogous sense so does the assembly of life forms themselves. I do not mean that humans create life, but that life creates life. All life arises from preexisting life. Remember? The source of the missing entailment is nothing other than the mother or progenitor cell.

It is not the cell's own proteins, its own DNA, but those of the mother cell that are entailing. Nature has provided a remarkable solution. The parent cell provides what is needed to specify location and sequence, as well as mechanisms to ensure proper compliance. It provides both the blueprint and the hands to build the structure. The mechanisms of timing and spatial disposition are the same as those the mother cell uses to *preserve* its own order. When the cell divides, all of the information and machinery is in place and is passed on to the progeny. Entailment is transgenerational. The "other" is the parent cell. That is how the cut is healed, how the salamander adds a new limb, and how new capillaries are formed. Cell division.

EPISTEMON *(his face turning crimson):* This changes nothing. As you point out, it is the *DNA* and the *proteins* of the parent cell that are entailing. Your "other" is nothing more than the material contents of the parent cell. It is not beyond or outside them. The strong-microreductionist principle can still explain everything that needs explaining. Proteins entail life, and DNA entails proteins. All we need is their presence in the mother cell. Our simple solution remains secure however much smoke you blow.

The Origin of Life

EUDOXUS: And then what? So, the missing entailment comes from the parent cell. And the parent cell is entailed by its parent, and so on back through history to the dawn of life. It is here, at the origin of life, almost 4 billion years ago, in the long distant past, that we must ask about self-assembly and entailments. How were these entailments brought into being in the first place?

Before I try to answer this question, we should understand what this excursion back in time has done to our project and on what shifting sands it has placed us. The ontologic problem of self-assembly that we have been discussing, which is accessible to laboratory research, has become epistemic and accessible only by examining the historical record. Direct scientific experimentation is no longer possible. The past remains in the shadows, however hardworking our historians may be.

Thanks to the work of geologists and paleontologists, we have a remarkable fossil record of life that has proven invaluable. Nonetheless, however remarkable and however helpful, that record remains extremely fragmentary. More important, as we delve back toward life's origin itself, there is no fossil record, not even a fragmentary one. Life's origin is not only in the shadows, shrouded in mystery, but remains completely hidden, unseen and seemingly unseeable by humans. Simply said, no matter how many hypotheses we devise, we have no evidence, no substantive knowledge of life's origin. How can we talk knowledgeably of entailments in such murky waters, lacking even an historical record as a guide?

And if this were not enough, even though it is a central belief of modern biology proved by no less a scientist than Pasteur that life cannot be generated spontaneously, the only way that we can account for its origin is by proposing just that. We have to propose that the spontaneous generation of life happened just once in the 5-billion-or-so-year history of this planet as the formative event of biology. A singularity.

Lessons from the Living Cell

Not only is the emergence of life often thought to be an occurrence so rare that it happened only once successfully, 4 billion years ago, its occurrence appears to be at tremendous odds with the great disordering force of the universe—the second law of thermodynamics. The second law tells us that disorder or entropy increases with time. As a result, we inevitably and inexorably come to a state of total disorder or equilibrium. That is where all matter in the universe is heading. The genesis of living organisms, and even more so their evolution into complex forms, exists in stark and seemingly contradictory contrast to the second law.

It seems that at some moment in the past, particular molecules were separated from their environment due to unknown events and against seemingly impressive odds, instead of being distributed more evenly as time passed, a far more likely happening. It was as if all of the oxygen in a room found itself in one corner (hopefully the one you are standing in) simply as a result of chance. Such a thing could happen, but it is not a possibility that should cause you sleepless nights. But in biology, it seems, it did happen. The system seems to have worked in reverse—order and organization increased over time, and the improbable became fact.

Thus, it is in a world of extremely unlikely events, at a time long past, that we must search for the missing entailments. How much confidence can we have in conclusions about entailments drawn from events about which we know next to nothing, that are both remote in time and seemingly rare? Can our conjectures ever be more than unsubstantiated speculations?

Epistemon and Eudoxus

Epistemon Views Life's Origin

EPISTEMON: I agree. But I am less glum about our circumstances than you. As you know, this lack of direct knowledge has not stopped scientists from thinking about the origin of life. Indeed, the subject has and will no doubt continue to attract much interest, as it has from the dawn of history. As for experimental science, we are not totally without possibilities. Although your research program admittedly failed, it showed that it is possible to do research today that bears on what happened at the time of life's beginnings. And some less overarching studies have actually enriched our understanding. Indeed, it has been demonstrated that a variety of organic molecules—including amino acids, the building blocks of proteins—can be derived from simple substrates that might have been available under conditions that might have occurred in the prebiotic world. And I am confident that future studies along the lines of your own will eventually lead to a fuller understanding of self-assembly as it first occurred. Still, when all is said and done, we are back to self-assembly, are we not?

And although it is difficult to obtain experimental data about this event in the far distant past, we can still use our faculty of reason. From what we know today about biological systems and about the physical and chemical world, we can evaluate possible scenarios for life's origin long ago. For example, we can be quite confident that today's organisms are descendants of those in the fossil record, and are a part of a common thread that passes back to life's beginnings. Also, the fossil record makes it clear that life first began with the simplest of organisms and that more complex forms evolved slowly over eons of geological time. And we also know that for all of their diversity, life forms share a common chemistry—the chemistry of nucleic acids and proteins. Moreover, we have no reason to think that the laws that governed the physical world at the time of life's origin were different from those that govern us today, and we can derive some grounded notions about what the prebiotic environment must have been like from geology and astrophysics. Although such extrapolations certainly do not allow us to determine the exact circumstances and character of first life, they do help sort things out.

Indeed, even though we have yet to be able to reproduce life's origin in our laboratories, or examine an historical record of the episode itself, we can deduce at least in general terms some of the central events that must have taken place. To begin with, we can conclude that the earliest life forms probably contained, or may even have been limited to, self-replicating molecules. Today, when life is often equated with the replication of DNA, perhaps this is self-evident. Without molecular replication, there can be no reproduction, and reproduction is needed to propagate life over time.

Also, in the beginning, life must have been in the form of aggregates of the replicating molecules, as well as perhaps other molecules produced by them or accumulated from the environment. Whether this occurred as the result of the direct formation of molecular aggregates or the partitioning of material in membranes or some other physical enclosure, we cannot say. But one or the other, or both, must have occurred. And this is so by definition. The first requirement for life is that it exist as an entity separate from its environment. Without the aggregation of matter, there cannot be life, at least life as we understand it.

No magic need be envisioned to fulfill these two requirements, just very ordinary chemistry. Even though present-day DNA is not, or at least has not been shown to be, capable of simple self-replication without the aid of various protein catalysts, the original substance—ancient DNA or its precursor, perhaps RNA or proteins themselves—must have been, even if only at a very low rate. And even though we have no specific knowledge of how this early replication took place and we may well never have such knowledge, it is clear that the process was a consequence of the ordinary chemical and physical properties of the replicating molecules themselves. *That is to say, Eudoxus, the origin of life was self-entailing.*

As for aggregation, we know that eventually membranes enclosed the contents of biological cells, though it is not clear how early this development came to pass. And direct molecular aggregation is among the most common physical-chemical effects. All molecules aggregate or precipitate under the right circumstances. As we have already discussed, given the right conditions molecules will self-assemble and, as a consequence, separate from the medium in which they are initially suspended. The molecules central to life, such as nucleic acids and proteins, as well as the lipid molecules that make up the membranes that enclose the cell and many of its internal structures, have a tendency to self-associate under environmental conditions that are quite ordinary on this planet.

In a simple system, molecules dissolved in a solution (solute molecules) have the potential to aggregate with others of their own kind also dissolved in the same medium (the solvent water). Unless they are inert, solute molecules interact with—that is, share electrons with—neighboring molecules in particular explicit molecular pairings. As such, even liquid mixtures display some minimal order. The interactions that occur depend on two central variables: the relative concentrations of the solvent and solute, and the relative strength of bonding of the different potential pairings (solute-solute, solute-solvent, and solvent-solvent). As the concentration of solute molecules is increased, they show a heightened tendency to associate with each other, rather than with the solvent. This, in turn, increases their propensity to separate physically from the solution. Such order-enhancing self-association can occur even when the aggregating molecules are present at quite low concentrations, and it is perfectly compatible with the second law of thermodynamics. The system is driven by the molecules' desire to seek their state of lowest free energy. When the solute molecules are present at a high enough concentration, this need is met by their association with their own kind. When this inclination becomes sufficiently great, the solute comes out of solution. It precipitates or crystallizes.

This is only freshman chemistry, but life's origin was a consequence of just such simple and common events. And it was the physical and chemical properties of the relevant molecules, and those properties alone, that determined the whys and wherefores of life's original self-assembly. This is all the entailment there was, and it was reductive self-entailment after all. Indeed, I see no other reasonable alternative, no other possible cause aside from metaphysics or life being brought here on a comet.

Whether likely or not, spontaneous generation happened successfully once, somewhere, sometime, as processes of chemical replication and molecular aggregation gave rise to the mother of all cells. This is only as it must have been. Replication was needed to make new cells; aggregation was necessary to have cells at all; and finally and crucially, self-assembly was the only way that aggregation could occur. It is true that we cannot say whether these very early "cells" were enclosed by a membrane or other physical barrier that further separated their biomolecules from the environment, or indeed served as the original means of aggregation, though we know that this eventually came to pass. Nor do we know what the particular molecular constituents of early life might have been. But whatever the singular details of the first cell's incarnation, the molecules that composed it had to include those that could replicate themselves and could self-assemble.

Epistemon on Spatial and Timing Constraints

EPISTEMON *(continuing):* This is all well and good so far, but I have yet to answer the question that so concerned you. How did spatial and timing constraints come about? As I have described it, first life lacked them. The first life form was probably not able to

Lessons from the Living Cell

specify the location or timing of interactions between its constituent molecules. But this should not be surprising. Most likely, it was not necessary for such simple systems, whose activities were probably little more than replication itself. At some point spatial and timing entailments emerged, most importantly, for the expression of protein molecules. But these entailments, like the aggregative and replicative effects that first produced life, are to be found in the structure of the molecules of life themselves.

I understand why you might think otherwise—why you think that an external referent is necessary. As you said, how could a particular spatial distribution of a molecule occur in a system that has no specified spatial symmetry to begin with? There appears to be a chicken-and-egg problem. Without the presence of a molecule at a particular spatial coordinate in the first instance, how can we introduce spatial symmetries into the system?

But this is not as overwhelming a problem as it might seem. For the most part, the specification of spatial relations came about as a consequence of the simple aggregative behavior we have just discussed. As molecular variety increased, there was a tendency of certain molecules to preferentially associate with each other. As you mentioned, the cell does not provide Newtonian coordinates for its contents, but things are organized in affinity groupings. There are nuclear proteins, mitochondrial proteins, and so forth. At the beginning, molecular groupings of this sort served as an important source of spatial entailment.

But as I am sure you realize, this would not have been sufficient in and of itself. As distinct

Epistemon and Eudoxus

aggregates emerged, they would have to be "tethered" to each other in some fashion to maintain or develop their "cellularity." For example, they might be trapped together within an enclosing membrane. But even this would not be enough. As you have pointed out, specific molecules and structures can be asymmetrically disposed *within* the cell. To use one of your examples, proteins are often asymmetrically distributed in membranes, being at one or another particular location along the cell's perimeter.

There are simple molecular solutions for this problem. For instance, a linear polymer to which different molecules or groupings of molecules can bind preferentially at one or the other end, or at various points along its length, could serve as the means of specifying location, or at least location relative to other molecules and structures appended to the polymer. As said, we know today that the cell's interior is crisscrossed by many such interlacing linear polymers, microtubules, microfibrils, and the like, that attach to membranes and to which a variety of other structures and substances may attach. This is one possible means of organizing the cell's contents in space. Another is planar organization. An example is found in the mitochondrion. In animal cells, the proteins (enzymes) of oxidative energy metabolism (oxidative phosphorylation) are in the mitochondrion. Many are found in and on its membranes in a particular spatial array that is designed to transfer electrons from protein to protein in a particular sequence. Here, the two-dimensional membrane provides the nexus for spatial organization.

A molecular solution can also be imagined for the problem of timing entailments. As you have asked, if the interaction between molecules in the protocell occurred in no particular order, then based on what sequence of events could time-ordered circumstances be imposed? This also is not an overwhelming difficulty. For example, it is likely that as different reactions emerged among molecules, they would occur at different rates simply by chance. As the mixture became increasingly complex, and there were more and more interactions, one or another molecule—substrate, product, catalyst, and so forth—would inhibit or accelerate this or that reaction also by chance. Similarly for binding interactions between molecules, a particular interaction might be favored, or only allowed, if preceded by another chemical interaction or reaction.

Eventually, the presence of such substances became a means of regulating the rate of protein-based chemical reactions as well as binding interactions that involved protein molecules. But equally important, over time an additional means of regulation emerged that *initiated* as well as modulated reactions. This was accomplished by the time-ordered production of particular protein molecules launched by the presence of other molecules, a process called induction.

In any event, it turns out that the parts do entail the whole, just as I have maintained, even if we have to include the parent cell as part of this whole. There is no need for external referents to explain the origin of spatial and timing constraints. And this is fortunate, because nothing else is available. There are no extramolecular causes. Large numbers of random structural alterations in DNA or its ancestors, passed down from parent to daughter cell, through generations extending over billions of years, gave rise to the modern state of affairs. These changes were responsible for the appearance of diverse protein molecules. And it was their properties, as well as those of the modified DNA, that gave rise to the features of modern-day cells that allow them to specify the location and timing of expression of their various protein molecules.

Mechanisms that are well known to us today emerged solely as a result of such molecular events. The timing of expression of various protein molecules involves chemical interactions with regulatory elements in DNA that produce a particular messenger RNA that in turn provides the code for the manufacture of specific proteins. And the ability to specify a protein's locus or site in or out of the cell can be found in its own structure as an amino acid code, as well as in structural features of receptor proteins at its various destinations.

In the final analysis, the evolutionary elaboration that took place was completely accounted for by mechanisms inherent to the molecules themselves—replication, aggregation by self-assembly, and centrally, alterations in the replicating molecules themselves. Whatever problems we may face in reproducing life's origin in a test tube, the evolution of spatial and timing constraints that have caused you so much concern were simply the result of structural changes in DNA and its product proteins. Nothing else need be imagined.

Eudoxus' Last Take and the Question of Adaptation

EUDOXUS: Yes, Epistemon, earliest life was probably the consequence of replication and aggregation without spatial and timing constraints. And those constraints must have come into being as a consequence of structural alterations, such as mutations, in DNA or precursors to DNA that gave rise to a multiplicity of products. But something important is missing from your analysis. You do not distinguish between the judgment that the emergence of life and biological variation depended on alterations in the structure of DNA, which is certainly true, and the similar, but quite different claim that these events were in and of themselves sufficient.

I know that as a strong microreductionist you cannot imagine what else could be involved beyond the molecules themselves. But as it turns out, the multitudinous alterations in the structure of DNA that were brought about by mutations, rearrangements, and error, and that produced the great diversity of protein molecules that we see today, did not in and of themselves give rise to life's varied and sundry properties. They were essential, prerequisite, but only the starting point. There is an immense chasm between this molecular variety and the extraordinary properties of life that arose consequently.

What is missing, what you have ignored or forgotten, is nothing less than the fundamental driving force of evolution—*natural selection*. It was by means of natural selection that the molecules you talk of became the material embodiment of the life forms that populate this planet. It was natural selection that connected them to life; that took their steel and cement and constructed life's edifice. And it is here that strong microreductionism ultimately fails. *It is in natural selection that we see that the parts do not entail the whole.*

As you know, all living things are constantly exposed to environmental challenges of one sort or another, and it is their respective fitness in the face of such challenges that determines whether they survive—whether or not they are selected by nature. Fitness can be defined as the organism's adaptability when confronted with particular environmental threats. And that adaptability is attributable to certain of the organism's features that, as such, we call *adaptations*.

While we have learned that natural selection is more subtle than this simple description, the fundamental principle still holds. The history of evolution is the history of biological adaptation to environmental necessity. Indeed, the life that we see today and that has existed historically, in all of its extraordinary diversity, can be viewed at a most fundamental level as a consequence of varied biological adaptations to the sum total of environmental exigencies that life forms have experienced down through the ages. That is to say, many, though certainly not all, of the distinctive properties of organisms that we see around us have evolved from antecedent elements as a result of their adaptive value in the process of selection.

Central to Darwin's genius was his realization that natural selection explained the remarkable, and previously inexplicable (except as a matter of faith),

Lessons from the Living Cell

variety of life. As a consequence, he provided a coherent explanation binding life's divergent forms together. But Darwin did not talk of natural selection in molecular terms. Neither he nor his contemporaries had substantial knowledge of the molecules and chemical reactions of life. Crucially, he knew nothing of the genetic code and the all-abiding role of proteins. Instead, he talked of natural selection as occurring to and between variant *organisms*, not molecules.

Was this merely a product of his times? Was Darwin unavoidably ignorant of the molecular basis of biological systems? Was his concept of selection among organisms merely a superficial description of the deeper, more fundamental molecular events that we understand today? Or was his description accurate? Can we only understand natural selection as acting upon whole organisms?

These are key questions. If the strong-microreductionist principle is to succeed, it must show that Darwin's concept of organismal selection is just a superficial description of deeper truths. If it is to justify its claim that the parts entirely account for the properties of the whole, then it must be able to reduce all of the many adaptations of life, the products of natural selection, to their component parts, their materiality, their molecular incarnations. If such a reduction can be made, I would submit to your view. It would indeed provide the deeper, more fundamental understanding you claim for the strong-microreductionist perspective. If it cannot, then adaptations are features of the whole, much as Darwin imagined them.

Epistemon and Eudoxus

At first glance, a reduction seems possible; indeed, it appears rather straightforward. After all, as we have discussed, molecular change—in particular, change in the structure of DNA—is responsible for the generation of the biological variants that are acted upon by natural selection. Changes in the structure of DNA and its constituent genes are reflected in changed protein products and, from time to time, in changed functional capacities of those products—and, accordingly, in changes in the properties of the organism. It is these properties, these features, that are acted on by natural selection. As a consequence, in this view, though natural selection seems to act on the organism—on its properties and adaptive features—since these properties and features can all be ascribed to the structure of particular proteins that are in turn the product of particular nucleotide sequences in DNA, in reality natural selection acts on DNA. If an organism does not survive a particular environmental challenge, then neither does its DNA. If it does, then so does its DNA.

If this conclusion is correct, then in one way or another we should be able to locate the adaptive features of the organism in its substituent molecules—in particular, in its function-producing proteins—either individually or as some specific summation of their features. It turns out that this is not possible for many adaptive features. Such adaptations exist only as they are found in the whole organism. They cannot be evinced in any sum of parts. Correspondingly, in going from DNA to the adaptive features of whole organisms, something is gained that exists only for the whole. Such features, wrought by the forces of natural selection, are, like certain numbers, nonreducible, noncompressible phenomena.

They are properties of the whole living system, and only the whole living system, whether cell or complex organism. And that living system is made whole, transcending its material necessity, in the sufficiency of these adaptive traits. Although these features are constructed, and constructed solely from life's material components, their adaptive essence is not to be found in the parts themselves, but in their interactions and relationships. As with all interactions and relationships, they exist as a consequence of spatial and timing entailments. It is in these entailments that we find the adaptive quality of life.

Indeed, Epistemon, more than that, it can be argued, and I believe convincingly, that it is these properties that actually make something living. That is to say, biological objects are not living because of their distinctive chemical ingredients, or because they replicate DNA or consume energy. They are alive because they display various unique adaptive capabilities. This is the "something more" of life.

Adaptation as Inherent

EUDOXUS *(continuing):* We can understand this better by considering geological evolution. As you know, it was contemporary ideas about geological evolution that prompted Darwin to think about biology in similar terms. Like biological evolution, geological evolution displays a natural selection. Oceans, seas, mountains, and valleys are affected by *their* environments. Oceans and seas evaporate,

Epistemon and Eudoxus

mountains erode, and valleys form. However, these events do not occur evenly for all bodies of water, all mountains, and all valleys. The rate and nature of these occurrences depend on local environmental conditions, but they also depend on the innate character of the bodies being acted upon. The salinity of the sea affects its rate of evaporation, and the structure of a hill—for example, made primarily of clay or granite—determines its rate of erosion. As a consequence of such material differences, a particular hill may survive an environmental challenge, while another may be washed away. All material objects undergo change in response to environmental forces; retaining, forming, and losing structure differentially. They grow and change, survive or disappear, and the rate and character of this evolution can be viewed as a natural selection among them. Because changes to different mountains and different valleys occur nonuniformly, we can think of their underlying structures as being *adaptations* to the forces of nature.

Such adaptations are due to inherent or "passive" properties of matter. By this I mean that they respond to the forces that act on them in accordance with their specific physical and chemical qualities. As such, all they can do is accommodate, acquiesce to the external forces imposed on them. Their passive compliance eventually leads to their transformation or the equilibrium state or state of maximum entropy or disorder, that is, to their demise.

Living beings, like all material objects, display such inherent adaptive properties. But if the animate and inanimate worlds were distinguished from each other only on the basis of such differences in their chemical and physical qualities, we would have no justification for concluding that mountains and rivers are inanimate objects, while plants and animals are living, any more than differences in the material substance and dynamic properties of water and clay are differences between living and nonliving things. This would squeeze life out of the living.

Adaptation as Transcendent

EUDOXUS *(persisting):* Horses and rocks do not differ from each other simply because of their different particular physical and chemical incarnations. They are objects of a different kind, and the difference between them is to be found in biology's unique evolution. It is here that Darwin's contribution was quintessential. He noted not only the similarities between geological and biological evolution, but the differences. He described how living objects adapt to environmental pressures in their own special fashion. Unlike inanimate objects, the adaptation of living things to environmental pressures can transcend their materiality. These adaptations are the unique features of life that distinguish the world of the living from that of the lifeless. Living beings can take action, do something, carry out particular maneuvers to their hoped-for advantage in the face of a particular environmental exigency. They are not limited to passive compliance.

And whether active or passive, such adaptations may not merely mitigate the organism's rate of attrition or change in the face of a particular external force, but ensure its survival whole, even totally unaltered. Remarkably, and crucial for the evolution of biological species, they may allow for the genesis of yet to be conceived progeny, of *organized things that do not yet exist*. The survival of one (the parent), permits (though it does not ensure) the future existence of the other (the offspring). Such adaptations are properties of the whole cell or organism. They are not reducible to their chemical essence.

Life's Nature

EUDOXUS: Consider the appearance of an animal. How it looks. It is well known that mere appearance, such as coloration or a fearsome demeanor, may be of crucial survival value, that is, can be an adaptive feature of an organism. Protective coloration may help an animal avoid a predator by making it all but invisible in particular surroundings. Or a fearsome appearance may ward off a potential attack on an otherwise vulnerable species.

Such features have meaning only in context. The object's coloration, for example, derives meaning from its juxtaposition to other objects of like cast. It is not the fact that some chemical or group of chemicals absorbs certain wavelengths of visible light preferentially that gives color its adaptive meaning, though that certainly occurs and is necessary for the effect. Rather, its meaning derives from the relationship between the color of the organism and the coloration of its environment. The adaptive property exists only for the whole organism. This can be easily understood with a reductio ad absurdum. If we could somehow remove the parts that display the particular adaptive coloration, reconstitute them as an object in its own right, and then place that object in an appropriate location, we would not expect it to display any kind of adaptation, certainly not any kind of biological adaptation. That would be silly. Protective coloration is a property of the whole; an irreducible, noncompressible property of life.

Similarly, in the Galapagos Darwin noticed that closely related birds (finches) had beaks with different shapes. Many years later David Lack traced these differences to differences in feeding opportunities particular to their microhabitats. The shape of the beaks seemed to be suited to what the particular bird ate and how it obtained its food. To explain such differences in strong-microreductionist terms, we would not only have to infer the beak's shape from its chemical and physical composition, but its adaptive purpose as well. This cannot be done. Nor can we tell whether a specific bone or series of bones makes a particular species swift, or powerful, or able to fly, simply from knowledge of the molecular constituents of bone. The substances that make up beaks, bones, and like structures, whatever their shapes and density, are essentially the same.

We might imagine that light bones are suited to flight. But this would only be an hypothesis, unless we otherwise knew that creatures with light bones actually flew. That is, we can determine the function of a particular design or structure, whether to aid in flying, running, or supporting loads, only if we have sufficient knowledge of the whole animal, or at least the whole functioning system within the animal. This is not just some sort of inconvenience; it is fundamental. However good we are at guessing, being blind to the external world of whole macroscopic beings, we cannot know whether a particular structure serves an adaptive function, much less what that function might be.

To better understand the importance of wholeness and context, and the insufficiency of materiality, as a last example, let us consider what is perhaps the most intuitively understood illustration of adaptive behavior—the fight-or-flight reaction of humans and other mammals. In this reaction, the animal either stands its ground or flees to a safe haven in face of a perceived danger. The reaction involves a massive reordering of the body's resources. The actions of many of its proteins and other biological chemicals, and most of its cells, tissues, and organs, are directed toward a single objective—survival in the face of peril. For example, a wide range of proteins are needed for the manufacture and secretion of epinephrine and the various steroids that are mobilized to enable the metabolic changes and enhanced muscle contraction that accompany the response.

But however many proteins and other chemicals may be called upon, and whatever cells, tissues, and organs may be involved, it is the organism and only the organism that displays the adaptive property. Yes, substances are manufactured and secreted, metabolic substrate is mobilized, muscles contract, and so forth. And yes, stronger or quicker muscles, more efficient metabolic pathways, hormones or input from the nervous system that more effectively mobilizes the blood circulation to the needed tissue beds, and so forth and so on will make an individual or species more likely to survive a dangerous encounter. But it is the organism, and only the organism, that displays the response and engages in or retreats from battle. The phenomenon, the adaptation, does not exist in its absence, no matter what the deep mechanisms and processes are that underlie it.

Even if we knew fully what all of the various molecules did under these particular circumstances—what the particular cells, tissues, and organs did, and exactly how they all interacted with each other to produce the response—we would still not know whether the person had stood his or her ground and fought or had chosen to escape from the scene, or even that such an event was occurring. Maybe the fear was imagined, or maybe the person was exercising, or maybe the person was imagining a dangerous encounter during exercise, or maybe the exercise was dangerous. The bodily functions would be essentially the same in any case. Heart rate would be elevated, cardiac output would be increased, respiratory rate would be increased, more oxygen would be taken into the body, more carbon dioxide would be exhaled, the blood supply to muscle would be increased and that to the gut decreased, kidney function would be diminished, thoughts of the philosophy of science would disappear, and on and on. The nature of the adaptation itself, however, to run or fight, cannot be explained in these terms.

Biological adaptations in their most highly developed form do far more than merely allow an organism to react to immediate environmental pressures, such as engaging or escaping a perceived threat. This is easiest for us to understand in our own species. For example, we may preempt the danger of environmental change by going to Florida for the winter. Or we can farm the land to provide food for ourselves to make us less vulnerable to the quixotic nature of an untamed environment. Or we can even change the environment in all sorts of ways that we call "development." Amazingly, however adequately or inadequately, we humans can plan for the future, and indeed try very hard to outguess nature in order to improve our prospects. These are adaptive actions that we say we undertake voluntarily. They are the "free will" of philosophers. Epistemon, I wish you luck in trying to reduce such adaptations to mere molecules and reactions.

This last comment essentially ended Eudoxus and Epistemon's already fragile collegiality. Now when they pass in the hall, they are polite; they talk of the day's events and institutional politics, but not about what concerns them most—the nature and mechanisms of life. Because Epistemon continued to believe that life could be reduced to molecules and reactions but could not prove it to the unconvinced Eudoxus, further conversation seemed pointless.

Immanence and Emergence

If this dialogue between Eudoxus and Epistemon has convinced you of what life is not—a mere compilation of its material elements—that is a great deal. But what about the question we began with, the question that embodies biology's quest: What is it that makes something living? Can one envision an affirmative explanation of life beyond the mere explication of its parts? Can we meet Epistemon's challenge to Eudoxus for an alternative that is consistent with our understanding of the material world?

The central dogma of modern biology is persuasive that life's essence is *immanent,* an inherent property of matter. It provides powerful evidence for this viewpoint in the genetic code. As we have discussed, at least superficially it appears that the genetic code provides all of the information necessary to produce life in the structure of its proteins, and since this code is a property of the DNA molecule and only that molecule, it can be said that life is immanent to DNA. In this view, life is not to be found in the irreducible qualities of the whole organism, or even in the properties of its various parts, but in this one central molecule.

Eudoxus claimed that this view is mistaken. The missing entailment, he said, is to be found in our distinctive evolution. He argued that not only do life forms depend on their unique adaptations, not only was it as a consequence of their presence that life evolved as it did, but that it was these adaptations that actually brought life as we understand it into being. More than that, Eudoxus maintained that these adaptive features defined life—what it means to be living. According to Eudoxus, a living entity is a material body that has the ability to adapt to environmental exigencies in this unique fashion, including the ability to take action. And life in toto, in all its forms and incarnations over the eons, can be seen as the sum of all such adaptations that have evolved and that have ever existed.

In this view, biological adaptations and hence life itself has an existence beyond, if necessarily dependent on, the goods that allow for its occurrence. As such, it is an *emergent,* rather than an immanent or inherent, property of matter. Although it arises from the material world, it cannot be reduced to it and as such is the noncompressible embodiment of life. As Eudoxus argued, DNA or no DNA, genetic code or no genetic code, there would be no life, could be no living beings, without the particular properties of the whole organism that emerge from their material incarnation to provide life's unique adaptations. They are what makes an object living.

CHAPTER 6

The Reductionist Experimental Program

You may seek it with thimbles—and seek it with care;
You may hunt it with forks and hope;
You may threaten its life with a railway-share;
You may charm it with smiles and soap!

Lewis Carroll, *The Hunting of the Snark*

And so the strong-microreductionist principle that the parts entail the whole does not apply to living objects. But this is hardly the end of the matter. And in a sense it is only the beginning. In many areas of biological research, the strong version of microreductionism has come to dominate practice in the laboratory, from the conception of experiments to their execution and interpretation. Strong microreductionism is a serious practical as well as theoretical problem.

As an experimental program, strong microreduction instructs us that it is through the in-depth study of the material components of life—its molecules, reactions, the substructures of cells such as the nucleus and mitochondrion, and the cells themselves—that we can come to fully understand life's nature. It is important to remember to distinguish this claim from that of weak microreduction, which simply holds that smaller, simpler, more fundamental parts underlie larger structures, and that we can learn much about the larger structure by investigating its underlying elements. The latter is a powerful research tool, the sharp edge of the microreductionist's sword, while the former is the sword's rough and shoddy edge, and as we shall see, leads to much confusion between what is actually known and what is merely hypothesis.

The Strong-Microreductionist Research Program

If asked, many experimental biologists would accede to the following tenets:

1. Life is a complex chemical system based on the unique properties of DNA and its encoded protein molecules.
2. Life's nature can be understood at the deepest level through the study of its chemical and anatomical constituents.
3. This applies to all of its features.

Though not often appreciated as such, these statements represent an endorsement of the philosophy of strong microreduction. I have heard them expressed by colleagues and students many times over a period of some 40 years, both publicly and privately. Though I cannot say how many are self-avowed strong microreductionists and how many are inadvertent believers, these three statements reflect an extremely common assessment among experimental biologists of life's nature and how to learn about it. We could take a

The Reductionist Experimental Program

poll, I suppose, but it might not be all that useful. We can't expect all biologists to have fully thought through their attitudes, or to be wholly honest about them if they have. Certainly we would learn something by asking, but a person who harbors a particular prejudice may not readily admit it, or for that matter even be aware of it.

But there are other, more useful ways to assess attitude. We can look at what people actually do, as opposed to what they profess. Do individuals function as *if* they hold a particular point of view; do they *practice* it, whatever they claim? We could ask: What is being done in the laboratory? What is the subject of published research? Where does the National Institutes of Health's (NIH's) money go? What topics are considered prestigious or hot, the subject of symposia, press reports, articles in *Science* and *Nature,* lead to nominations for the Nobel prize? What graduate programs are thriving, and which have disappeared; and what courses are graduate students in biology required to take, and which are elective? Such information would provide a more accurate account of beliefs than an opinion poll.

When one seeks to answer the question in this way, the conclusion is inescapable. Any survey of National Institutes of Health grants, of the scientific literature, or of graduate education leads to the conclusion that the strong-microreductionist perspective has been and remains paramount. Students today are often well trained in molecular subjects, but only superficially educated in other aspects of biology. Soon, I suspect, even physicians will know more about genes than about the physiology of the heart.

The main emphasis in experimental biology during the twentieth century, increasingly so as the years passed, was on learning about life by rendering its material constituents intelligible. The subjects of greatest prominence have been the characterization of the chemical components of living things, uncovering their reactions, and identifying the microscopic structures which they form and in which they are found.

The Human Genome Project, the first physicslike megaproject in biology, is the culmination of this research program and perhaps its ultimate achievement. We have spent billions of dollars—more than has been spent in many areas of biology taken together over the years—to establish the chemical sequence, the whole sequence, and nothing but the sequence, of human DNA. As already noted, this task was not undertaken merely to satisfy the desires of a group of biologists who were curious about the complete chemical structure of this molecule. It was argued that this undertaking was crucial if we hoped to understand life's nature at the most fundamental level, the level of its genetic chemistry.

The Happiness Workers

There is what has become a generic experiment today in which a specific gene is found that is associated with a complex property of the whole organism. On detecting the gene, an announcement is made that the gene for that particular property—say, "happiness," to take a seemingly extreme example—has been discovered. The claim is not merely that this property is in some unknown fashion associated with the particular gene, but that it is *attributable* to it and its protein product. The evidence for drawing this conclusion is that when the gene is absent, or is present in one or another altered form, the particular property is no longer seen. For example, the animal no longer displays activities associated with happiness. It may be inactive, lack appetite, have lost its sexual drive, and so forth.

The Reductionist Experimental Program

Such inferences from gene to whole organism are typical of the strong microreductionist's confusion. Of course, the "happiness workers" have not demonstrated that their gene is actually responsible for happiness. What they have shown is that it is necessary for it to be manifest, assuming that their measurements of the property are appropriate. What they have done is infer *sufficiency* while only having demonstrated *necessity*. This is like believing that electricity produces the *content* of television programs, because that content disappears when we pull the plug. This kind of reasoning is typical of the strong-microreductionist mind-set.

The Essential Confusion

At least some of this way of thinking has been inadvertent or unintentional, a natural product of the way in which we design and carry out experiments. Studies in experimental biology often begin with the unassailable view that there is much to be learned about life by studying its parts—weak microreductionism. With this understanding in mind, investigators set out to do just that. But once they have carried out their experiments on molecules, reactions, cellular structures, or even on cells themselves, they feel obliged to explain what they have learned not merely in terms of the parts they have studied, but in reference to the functioning whole. This is not surprising. Such explanations are, after all, the goal of all experimental biology.

But all that the investigators can legitimately do in this circumstance is imagine what the relationship between their results and the unexplored whole might be. In the absence of the necessary knowledge about the whole functioning system, such extrapolations cannot be made in a logically acceptable fashion. As we shall see, to attempt to make them is to accept, however tacitly, the strong-microreductionist prescription.

Whether it is a conscious belief or not, the expectation of much of modern experimental biology is that if we learn about the parts of life systems, particularly genes and proteins, in great depth and with great rigor, we can in our wisdom infer the nature of the whole living system without ever actually having to examine it. Research that relies on such thinking is fraught with the risk of misunderstanding and misperceiving both nature itself and the extent of our mastery of its mechanisms.

Perhaps the problem this creates can be appreciated in an exercise I occasionally give students in seminars and classes. I ask them to try to distinguish between the *results* of a piece of published research on one or another part of a particular biological system and the *interpretation* given the results. The student usually thinks that this is an easy, even a frivolous, assignment. This is particularly so in biology, where research papers are usually organized into clearly named Results and Discussion sections. Obviously, they will find the results in the Results section and the discussion of their possible meaning in the Discussion section. Where else? Consequently, they are often dismayed when I point out that they have confused the two when they present their analysis to the class. The authors, they come to realize, have amalgamated theories, models, assumptions, and evidence into one large *Gemisch* labeled "Results." When this is done, ambiguous data often seem to provide powerful proof for a particular theory or point of view because data, theory, models, and assumptions have all become one and the same—results. As we shall see, that this seems a justifiable way to present material is an outcome of the sway that the strong-microreductionist way of thinking has over us.

The Reductionist Experimental Program
Simple Systems

Whatever problems physics confronts in attempting to apply the strong-microreductionist principle, they pale in comparison to those in biology. The experimental systems in physics are extremely simple, and for all intents and purposes (quantum and Heisenberg uncertainties aside) can be fully defined or specified. In experimental biology, the situation is far more difficult. With the exception of the simplest chemical (structural or reactive) systems, we confront complexities of a nature and extent that do not exist in the inanimate world of the physical sciences. Most important, these are complexities that usually cannot be fully characterized. Although biologists control their studies as best they can, they often find themselves working with an unknown number of unspecified variables. It is these unspecified variables that are responsible for the all too common inconstancy of biological data, even well-controlled laboratory data. And it is for this reason that experimental biologists often resort to statistical validation for their observations, or ignore this need at great risk.

Biologists often try to avoid, or at least alleviate, some of this variability and its attendant uncertainties by restricting their efforts to the study of biological molecules and reactions in isolation, or to parts of cells, or even to whole cells of multicellular organisms in tissue culture. They believe that by studying less variable and more easily characterized molecular, subcellular, and cellular events they can come nearer to the simplicity of research in physics—or if not physics then at least chemistry. And this is most certainly true and has been a very useful strategy.

But in accepting this strategy, many come to accept the far more questionable proposition that nothing about the whole system, or at least nothing of importance, is lost in the process. It is this judgment that is in question here. The belief that the whole, whether cell, tissue, organ, or organism, can be entirely characterized inferentially from a deep and complete mastery of its underlying parts is not only deeply flawed, but can lead us away from, not closer to the truth, even when the opposite seems to be the case.

CHAPTER 7

A Real-Life Parable

*Not chaos-like together crush'd and bruis'd,
But, as the world, harmoniously confus'd:
Where order in variety we see,
And where, though all things differ, all agree.*

Alexander Pope

The conflict between the strong microreductionists and those who believe that the material substance of living things does not entail everything is well over 100 years old, and many important discoveries in modern biology have concerned this difference in perspective. Controversy has arisen, particularly when different groups have studied the same process from these incommensurate points of view simultaneously. I will give two examples of this in the following chapters. The first goes back to the years just before, during, and after World War II, and the second is somewhat more recent.

My intention is to show why strong microreductionism is bound to fail as an experimental strategy, and how, paradoxically, its adherents are often convinced that their understanding of nature is greater than it actually is. I will consider these issues for two essential, but seemingly quite different, biological processes. The first is the mechanism that is responsible for the development of tension in muscle, that is, muscle contraction. The second is how chemical substances, particularly proteins, are moved both within and out of their cells of origin; that is, their secretion. First, about muscle contraction.

The mechanism that underlies the contraction of muscle, particularly the skeletally attached muscles responsible for locomotion, has long been a subject of great interest. There are worthy practical reasons for this: to help in the treatment and cure of various diseases in which muscle activity is depressed or weakened. But historical interest in how muscles contract has been less a reflection of this estimable goal than the longing to grasp the fundamental nature of life. Muscles play the central role in motion in complex metazoans, such as humans, and life has traditionally been associated with motion and death with its absence. At the time of the ancient Greeks, it was thought that to understand the nature of muscle activity was to understand motion in living things, and to understand motion was to understand life.

Galen and Galvani

It was with this goal in mind that in the second century A.D., Galen, a physician and arguably the first experimental biologist, tried to determine the underlying cause of muscle contraction. He observed that a single large tube or vessel was attached to each muscle. We know today that this structure is the nerve trunk that travels from the spinal cord to the muscle it innervates. Its presence suggested to him that the force that muscles generate during contraction might come from this extrinsic source. He surmised that the vessel might carry substances that cause contraction, and attempted to distinguish between this hypothesis and the intrinsic causation of contraction, from within the muscle itself, with a simple experiment.

A Real-Life Parable

He reasoned that because tension seemed to be generated throughout the whole muscle, if the mechanism was intrinsic, it too would be distributed all through the muscle mass. In this case, if one were to cut the muscle in half, each half would retain its ability to contract. On the other hand, if the tubule were responsible for contraction, cutting the muscle in half would disconnect one half from it, leaving it flaccid, while the attached half would still be capable of contraction.

What did Galen find? Even as he was cutting the muscle, both halves began contracting toward their insertion on the bone. And when the muscle was totally transected, both parts retained their contractile ability. This demonstrated to his satisfaction that the contraction of muscle was due to mechanisms intrinsic to its structure, and not to the delivery of contractile substances via the attached vessel.

But Galen's experiments did not end the matter. Although his conclusion carried great weight, the question of whether muscle contraction was due to intrinsic or extrinsic causes was asked by many scholars down through the centuries. Indeed, almost two millennia later the idea that contraction was due to an *extrinsic* cause received a great boost. In the early nineteenth century, the Italian scientist Luigi Galvani was interested in the nature of electricity and had begun studying its expression in living things. He focused particular attention on muscle and Galen's vessel. Using a muscle from the limb of a frog, he stimulated the attached vessel with an electric current. The muscle contracted. Moreover, if he placed the vessel from one muscle on the surface of another muscle and electrically stimulated the vessel, the second muscle contracted as well. Contrary to Galen's conclusion, this observation appeared to provide a clear demonstration that the attached vessels were responsible for muscle contraction.

Not only that, but Galvani had discovered the causative "substance" within the tubules—electricity.

Today, as the result of work by many scientists in many fields, we understand the distinction between extrinsic and intrinsic causes of action in biology quite well. And this understanding pertains not only to muscle contraction and other kinds of motion or motor activity, but quite generally to the "irritability" of living things—that is, to their reactiveness or responsiveness to extrinsic stimuli and causes in general. Not surprisingly, there are both intrinsic and extrinsic causes of muscle contraction and all other biological phenomena. The force of contraction is generated by mechanisms intrinsic to the muscle, as Galen thought, but Galvani's electrical impulses, traveling along the nerve trunk, account for the muscle's ability to respond to external stimuli.

We will see how all of this comes together in a bit. But first, with Galen's and Galvani's experiments in mind, let me describe attempts more than 50 years ago by two twentieth-century scientists and their students to understand the mechanism of muscle contraction. Even though both scientists had the same goal, their *philosophical* perspectives differed greatly—indeed, seemed contradictory. As a consequence, their research programs were entirely different in character and conception. As we shall see, this led them to very different explanations for how muscles contract. The strategy of one of the scientists, Albert Szent-Györgyi, was an application of the strong-microreductionist principle, while that of the other, Lewis Heilbrunn, represented its vociferous rejection.

Szent-Györgyi

The story begins with Albert Szent-Györgyi. Szent-Györgyi was a brilliant biochemist and a recent recipient of the Nobel

A Real-Life Parable

prize for his discovery of vitamin C. He had just set up shop at the Woods Hole Marine Biological Laboratory in Woods Hole, Massachusetts, after having emigrated from his native Hungary in the shadow of World War II. Szent-Györgyi had become interested in how muscles contract and had launched an active research program designed to discover the underlying mechanism.

Up to that point, contemporary work on muscle contraction had been centered on characterizing the mechanical properties of the intact process. For example, how much force did contraction generate? What were its kinetic parameters; how fast was the force generated, how fast did it decline, and how long did it last? And under what circumstances of state, such as the length of the muscle, its prior contractile condition, and so forth, did the properties of the process vary?

Some years before, a Nobel prize had been awarded to the British scientist A. V. Hill for his studies on muscle contraction. Hill had made remarkable measurements that correlated the kinetics (time course) of contraction to its energetics (the generation of heat). By developing a method to measure the tiny amounts of heat that muscle cells generate during a contraction—that is, prior to, during, and subsequent to the generation of tension—he was able to divide the process into a series of brief, defined stages. These thermal signs provided a valuable road map for the future discovery of the underlying biochemical and physical mechanisms.

Thus, by the time Szent-Györgyi became interested in muscle contraction there was a relatively well-developed energetic and general mechanical description of the process. But little was known about the underlying mechanism, and that was the prize he was after. As a strong microreductionist, he believed that if he could determine the chemical substances responsible for contraction, he could understand the mechanism of muscle action.

Actin, Myosin, and ATP

The first challenge was to extract and isolate the proper constituents, and to do that it was necessary to break open the muscle cell. Szent-Györgyi focused his attention on the central organic substances of muscle cells—their proteins. Two in particular, actin and myosin, were present in large quantities and hence seemed likely to play an important role in the contractile process. Following earlier work by others, and after some effort, his group was able to isolate them and characterize a number of their chemical and physical properties. But this was just the first step in his ambitious plan. Having isolated and characterized the proteins, he intended to reconstitute the contractile mechanism in a test tube.

What Szent-Györgyi found was striking and seemed to confirm his strong-microreductionist conviction about how to discover biological mechanisms. His strategy was bold and simple. When he added a chemical source of energy, adenosine triphosphate (ATP), to various extracts of the muscle cell, or to the isolated proteins themselves suspended as a viscous gel, the material contracted. He called this phenomenon *superprecipitation*. In some systems, as the ATP was broken down over time, the material reverted to its former relaxed state. This cycle of contraction and relaxation could be produced over and over again, simply by adding more ATP to the mixture. It was from observations such as these that Szent-Györgyi concluded that muscles contract by some kind of reversible congealing or coprecipitation of actin with myosin, fueled by ATP. When he reported his initial results, they were greeted with enthusiasm. It seemed that after millennia of

A Real-Life Parable

speculation, the mechanism of muscle contraction had been discovered, or at least was on the verge of discovery.

Beyond the particularities of his findings, Szent-Györgyi also seemed to have proven an important general principle: that a physiological process could be reconstituted in a test tube. The mechanism of muscle contraction could be understood simply by studying the properties of the relevant molecules in isolation. The whole cell was not needed; its complexity could be avoided, and yet the mechanism could be thoroughly mastered. Szent-Györgyi had apparently validated the strong-microreductionist perspective for biological systems.

He realized that much additional work had to be done before a comprehensive understanding could be achieved, but he was confident that this was just a matter of time and effort. Szent-Györgyi and numerous other investigators studied the properties of *superprecipitation* and similar phenomena over a period of years as they tried to tie down the molecular mechanism responsible for muscle contraction. When that task was complete, the mechanism that muscles use to generate force would be known with complete molecular specificity.

The strong-microreductionist principle seemed to have had a great experimental success—perhaps its first fully realized success. It seemed only a matter of time before there would be many more successes if scientists simply followed Szent-Györgyi's excellent strong-microreductionist example. As for muscle, it was only a matter of time before Albert Szent-Györgyi would be awarded his second Nobel prize.

Heilbrunn

At about that same time, at the University of Pennsylvania in Philadelphia, Lewis Heilbrunn was also attempting to understand how muscles contract. His approach differed from Szent-Györgyi's in fundamental ways. Unlike Szent-Györgyi, Heilbrunn did not study isolated chemicals. Indeed, he scoffed at the idea. He thought that trying to understand the mechanism of muscle contraction by homogenizing tissue and extracting the cell's chemicals was like attempting to understand how a car worked by grinding it up; isolating its material components, such as iron, copper, and so forth; and then looking for clues about its capacity for locomotion in the properties of these substances.

Heilbrunn believed that the secrets of cellular mechanism could be uncovered only by studying the whole cell and its properties. He was not a biochemist like Szent-Györgyi, but a general physiologist, and he emphasized the importance of learning about the properties of whole functioning cells and organisms as the route to deeper understanding. Heilbrunn and many other general physiologists were weak microreductionists. Even though they believed that the chemical composition of biological systems needed to be understood to gain complete understanding, they saw such knowledge as the biochemical cart to be drawn by the physiological horse. General physiologists held that comprehensive knowledge of the properties of the intact system was prerequisite to a functionally meaningful understanding of its chemistry. To put biochemical knowledge first was not merely to put the cart before the horse, getting it backwards, but would inevitably result in becoming lost in a *Gemisch* of chemicals of uncertain provenance.

According to Heilbrunn and the general physiologists, unless the intact process was understood, one could never know whether a proposed chemical mechanism accounted for its properties, because those properties, unstudied, would be unknown. For example, to validate his claim that he had discovered the mechanism of muscle contraction, Szent-Györgyi would somehow have to demonstrate that his proposed chemical mechanism generated force. Not only that, but exactly the same force as an intact muscle, mole for mole of contractile chemical. More broadly, he would have to show that the putative chemical mechanism could account for all of the known properties of the intact system, its particularities of rate, state, and energetics. Unless he could do that, his model would fail.

Protoplasm

As a young man, not yet out of his teens, Heilbrunn left his home in New York and went west to Chicago to enroll as a graduate student in zoology at the University of Chicago. In the early years of the twentieth century, the University of Chicago was a hotbed of reductionist general physiology. The young Heilbrunn was particularly fascinated by studies that attempted to understand the physical character of the cell's contents—in contemporary terms, its *protoplasm*.

When Robert Hooke first described cells in dry sponges more than 200 years earlier, he had characterized them as empty microscopic spaces of more or less polygonal geometry that were molded and formed by an enclosing web of fibrous elements. He called the spaces *cells* to liken them to the bare, almost empty living quarters of monks. Hooke, and subsequently many others, quite naturally looked to the surrounding web of fibers, not these empty spaces, for the living material. Life was in the fibers, not the cells. Even when it became clear that living cells were not empty like Hooke's dry sponge cells, their contents were thought to be nutrients, droplets of fat, glycogen, and the like, present to nourish the living fiber.

Lessons from the Living Cell

With the advent of cell theory, a critical gestalt shift took place. Life was now in the living cell, not the surrounding fiber. Today we know that the fibrous material thought for so long to be living consists for the most part of noncellular, microscopic fibers of protein (collagen) that criss-cross spaces between cells, the so-called connective tissue. But if life was in the cell, where and what was it exactly? That is, if life was intrinsic to cells as cell theory proposed, then to which of its parts could it be ascribed? What was the living material *within*?

In some of the earliest *invasive* studies on cells, when egg cells, protozoans, or other similar cells were punctured or otherwise broken open, a gelatinous substance that moved emerged. Although it stopped moving after a while, its movement suggested that it was the living substance. There was considerable disagreement about what this material was and what to call it. Eventually the term *protoplasm* stuck. If one wanted to understand the physical and chemical properties of life, then one had to understand this ooze, this moving protoplasm.

Heilbrunn and other general physiologists attempted such a characterization, not of the ooze after its isolation from the cell, but within it, in situ. Looking at whole single cells in the visible-light microscope, particularly protozoans and large egg cells like that of *Arbacia*, the protoplasm could be examined directly in the living cell. It seethed with motion. In the case of motile organisms such as the amoeba, this internal motion seemed to move the organism itself. In a series of studies, Heilbrunn was able to characterize the physical state of the protoplasm as a stable suspension of particles, some sort of colloid.

It seemed that in some as yet unclear fashion, its colloidal nature gave this material its lifelike quality, its motion. At the time, biological chemistry was still in its infancy, and the

A Real-Life Parable

chemical composition of the protoplasm was in great part unknown, or at least uncertain. Proteins were clearly important, but their structure and role were still a mystery. Still, whatever the chemical contents of the protoplasm might be, it was its physical properties that were most important to the general physiologist. After all, it was on these properties that life depended.

As more and more became known about the chemical contents of cells, the notion of a living protoplasmic substance became less and less realistic. In the final analysis, there was no credible evidence that there was some particular material substance inside a cell that was living. Protoplasmic theory was reasonable enough, given then-current knowledge, and it certainly was a parsimonious hypothesis, attributing the living state to a single (albeit poorly understood) substance, but it was clearly wrong.

As the biochemical contents of cells came to be better defined, the substance referred to as protoplasm, if it existed at all, seemed to be nothing more than a complex and quite variable mixture of chemicals with diverse properties suspended or dissolved in water. They were not living in themselves, either singly or as a mixture. The startling conclusion was that although the cell was alive, it appeared to contain no living substance! When one broke it open to recover its contents, life seemed to evaporate.

By the late 1930s the term *protoplasm* had been replaced by a new, less ambitious, but more realistic term—*cytoplasm*. Unlike protoplasm, cytoplasm was not considered a living material. It was the ground substance of the cell, the material in which large intracellular structures, such as the nucleus, were suspended. When the electron microscope was applied to biological systems, greatly increasing the resolution (magnification) at which cells could be examined visually, a variety of very small anatomically distinct structures were discovered that had not been seen before, at least not clearly. As a consequence, *cytoplasm* was defined even more restrictively. As more and more was learned about particulate structures within cells, the cytoplasm shrank, and came to represent a smaller and smaller part of the whole. Today we can say that the cytoplasm accounts for about 50 percent of the volume of the plant or animal cell on average. The biggest structures we attribute to it are various large protein polymers.

Thus, the idea that we might be able to further our understanding of life by reduction to a particular, more fundamental subcellular living substance failed. One could not point to any particular part of the cell and ascribe life to it. The cell seemed to be nothing more than a sac of biochemicals that were not in any observable sense living in themselves. For many, the notion of the living cell seemed to vanish. Biology was chemistry. The rest was illusion. To talk of a living cell was simply to confuse things and to be distracted from the important task of establishing the chemical composition of the *biological*, no longer the *living*, cell.

But for others this realization had quite the opposite meaning. For them it meant that the original holistic view of the cell as the irreducible unit of life had to be reinstated. It was only the cell as a whole that was living. To understand life we had to understand *its* properties, the properties of the whole. This was Heilbrunn's view. He believed that the properties that gave the cell life were those of the intact object. They could not be found in isolated bits or pieces one might extract from it. If one wished to grasp life's nature, its particular attributes and processes, one had to study it whole.

A Real-Life Parable
The Sol/Gel Theory of Motion

Heilbrunn was interested in how the flow of cytoplasm gave rise to the movement of motile cells from place to place. He believed that this occurred as the result of a cyclical transformation of the cytoplasm from a sol to a gel. That is, motion occurred due to a change of state. At any particular time certain parts of the cell were in a liquid (sol) state, while other parts were coagulated (gel). Heilbrunn thought that the cell's exterior layer, which was called the *cortex*, was coagulated, while its interior, or *medulla*, was fluid. The cytoplasm flowed when portions of the cortex became fluid and the interior simultaneously gelled. A process was envisioned in which a spatially directed transformation from gel to sol and back again to gel produced cytoplasmic flow that, in turn, gave rise to locomotion.

But what caused this transformation? It had to be initiated by something. Heilbrunn proposed that the calcium ion played this role. It was known that calcium was necessary for *motor activity*—that is, movement or change in shape—in many cells. Most famously, toward the end of the nineteenth century Sidney Ringer found that in its absence heart muscle ceased to contract.

Heilbrunn explained things as follows. In a quiescent motile cell, calcium was bound to the cortical gel, while the fluid medullary region was calcium poor. Not only did the calcium ion bind to the gel, it was responsible for the gelled state itself. That is to say, in its absence that region of the cytoplasm would become fluid. Cellular motion was initiated when calcium was released from the gel at the cell's periphery, and diffused into the calcium-poor central region.

As a consequence, the cortex became fluid, while the medullary region, momentarily containing substantial amounts of calcium, became gelatinous. Cytoplasm would flow as a result of this transformation and the cell would move. Even though notions of how cells move are far more specific today, our modern ideas are quite analogous to Heilbrunn's. Today we talk of the assembly and disassembly of filamentous protein polymers in different regions of the cell as generating movement, not sol/gel transformations. But like the gels, the polymers are formed by the extraction of soluble substances (protein monomers) from the cytoplasm to form an insoluble material.

Heilbrunn's View of Muscle Contraction

Heilbrunn thought that muscle contraction was due to a similar process. Indeed, he thought that all biological motion was at base due to the exact same process. In muscle cells, as in protozoans, the calcium ion would diffuse from the cell's periphery to its center, producing the same kind of changes. In this case, calcium would be mobilized by stimulation of the nerve to the muscle. As a result, it would diffuse to the central region of the cell where it would induce the gelled state, causing the muscle to contract. During muscular relaxation, the calcium ion would diffuse back to its storage site in the cell's cortex, and the medullary region would again become fluid.

Heilbrunn settled on an experiment to test his hypothesis and assigned it to a graduate student, Floyd Wiercinski. Like Szent-Györgyi's, Heilbrunn's experiment was simple and bold, except that it took place in an intact muscle cell, not a test tube. He reasoned that if calcium caused muscles to contract, then its injection into a flaccid muscle cell would lead to contraction. Wiercinski carried out the experiment by dissecting out small muscle fibrils from a frog's leg muscle, and then injecting small amounts of a calcium-containing fluid into the intact cells with a fine hypodermic needle and syringe.

What happened was striking. The muscle cell contracted immediately. When other substances, such as sodium chloride, were injected instead, it did not contract. The effect seemed to be calcium specific, and as such provided direct support for Heilbrunn's theory. The calcium that Wiercinski had injected into the cell seemed to be acting in lieu of the cell's endogenous calcium pool.

Two Views in Opposition

Szent-Györgyi had extracted the chemicals that he thought were involved in muscle contraction and examined their properties in vitro. Heilbrunn, on the other hand, studied the process in the whole muscle cell, seeking evidence for a putative calcium-induced sol/gel transformation. Szent-Györgyi's approach led him to the proteins actin and myosin, while Heilbrunn's implicated the calcium ion in the contractile process. Well then, who was correct? Did muscle contract because of the superprecipitation of actin and myosin or because of the mobilization of the calcium ion and a change in the physical properties of the cytoplasm?

Before I answer this question, let me briefly set the stage. Like Szent-Györgyi, Heilbrunn was also an active member of the Woods Hole laboratory. He had lived there earlier in his life and now spent every summer at his Cape Cod cottage on the pond adjacent to the laboratory. So there was much opportunity for the two men to discuss their conflicting ideas, and they did so from time to time. In the larger world of science, Szent-Györgyi's actin-myosin theory clearly dominated thinking, but at Woods Hole, and especially among Heilbrunn's acquaintances and students, controversy raged.

Lessons from the Living Cell

From Heilbrunn's point of view, all that Szent-Györgyi had shown was that proteins from muscle precipitated in a test tube. What relevance could this possibly have for understanding the mechanism of muscle contraction in whole muscle cells? To propose that what he had observed between isolated chemicals was identical to what happened in the intact cell was little more than unsubstantiated speculation. Szent-Györgyi had no real idea of what happened in the cell. Heilbrunn complained that one could learn little about the mechanism of muscle contraction by such "blenderizing" of cells.

For his part, Szent-Györgyi responded that how else could one learn about a mechanism except by examining its component parts? Without dealing with the underlying chemistry, little would be learned no matter how many experiments Wiercinski carried out. And whatever understanding Heilbrunn might achieve would be excruciatingly vague and incomplete. Szent-Györgyi considered Heilbrunn's experiments hopelessly naïve. How could one draw any conclusions about the mechanism of muscle contraction simply by injecting a substance into muscle to see if it contracted? Indeed, Szent-Györgyi and his colleagues—and indeed the community of scholars more generally—thought that Wiercinski's effect was probably an experimental artifact of some sort. To produce it he had injected relatively large amounts of calcium into the cell; hence, it probably had little to do with the real business of muscle contraction. Anyway, how could Heilbrunn know whether it did or did not?

Two Theories Merge

The two men started with what appeared to be mutually exclusive approaches, so it was not surprising that they obtained very dissimilar types of evidence. Nor was it sur-

A Real-Life Parable

prising that they drew quite different, and seemingly incompatible, conclusions. Heilbrunn and Szent-Györgyi never resolved their differences about the mechanism of muscle contraction, but others did. Neither has had it entirely his own way. They were both plainly wrong, but equally as plainly right. It became clear in the 1950s and early 1960s how their observations and theories mapped onto the process of muscle contraction as we presently understand it. As so often happens in science, clarification had to await the application of a new technology, in this case electron microscopy.

The electron microscope was invented by physicists in the 1930s to examine matter at the atomic level, but it was a failed curiosity. It turned out to be of little value in the physical sciences, for which particle accelerators and other methods of far greater resolving power were developed. And so even though the device underwent substantial development after its invention, it remained largely unutilized.

But this initial failure was not to be the technology's permanent fate. During and just after World War II, biologists were given access to several unused electron microscopes to assess their potential for biological research. The results were revolutionary, and biology found an important new experimental tool. Viruses were seen for the first time, and as the method was developed, the fine substructure of all manner of cells that had been previously hidden from view, or only seen dimly in the visible-light microscope, could now be observed, often with remarkable clarity. As a consequence, a whole new world of *ultrastructure* was exposed that has since been immensely influential in virtually every area of biology.

Obviously the electron microscope was a microreductionist device. It allowed the examination of smaller, and hence presumably more fundamental, structures than had hitherto been possible. It offered an important advantage over Szent-Györgyi's chemical microreduction, because the cell's components could be seen more or less in their natural location. The cell did not have to be homogenized for the scientist to look into the black box that was the cell.

Muscle contraction was one of the first physiological processes to be studied by electron microscopy, and it turned out to be a very fruitful marriage of problem and technique. But neither Szent-Györgyi's proteins nor Heilbrunn's calcium were the initial focus of interest. Rather, and quite naturally, a central goal of early electron microscopic studies in general, and those on muscle in particular, was to develop and validate the method. Could one make clear images? Could sufficient contrast be obtained to plainly distinguish different structures? And, most important, were the images accurate? That is, to what degree did they provide an accurate portrayal of things as they were in the intact cell, and conversely to what degree were the images distorted, or even wholly artifactual? Both the high-energy electrons that made up the light beam and the need for a substantial vacuum to allow them to penetrate the sample were very damaging to ordinary biological material. To make the method useful, beam- and vacuum-induced damage and destruction had to be prevented, or at least ameliorated. To be protected, the sample had to be altered. Indeed, before it could be viewed successfully, it had to be almost completely transformed.

First of all, all of the cell's water (about 75 percent of its mass) had to be removed and replaced with a substance that did not vaporize in the viewing chamber's vacuum. This replacement had to be accomplished perfectly, without introducing swelling, shrinkage, or any other distortion that might alter the object's appearance. Plastic materials such as acrylic and epoxy became the common substitutes. Furthermore, the sample had to be cut into extremely thin sections,

A Real-Life Parable

less than one-tenth the thickness of the average cell (or less than 100 nanometers, or nm) for a sufficient number of electrons to penetrate to form an image.

Furthermore, it was necessary to add substances to the sample to increase the contrast between different structures. Because most biological structures are composed of the same elements—mainly carbon, hydrogen, oxygen, and nitrogen—they absorb electrons in roughly equal amounts. That is to say, their density is roughly the same, and thus their relative lucency or opacity to the electron beam is also roughly the same. This made them hard to distinguish from each other. To make a structure more opaque than its neighbor, substances that strongly absorb electrons, such as heavy metals, that preferentially bind to it had to be added.

We will go over this method in more detail in Chapters 9 and 10, but suffice it to say here that these procedures, as well as the process of viewing itself, had the potential to introduce all kinds of artifacts into the image. Artifacts were not only possible, they were common, even usual. As a consequence, and not surprisingly, great effort was expended during the initial period of the method's development to avoid artifacts. Even today, in spite of the maturity of the method, artifacts are not uncommon features of electron micrographs.

But whatever problems persist today, the faithfulness of electron microscopic images to the natural object was a question of major concern during the initial development of the method to examine biological samples. Somehow the fidelity of the images had to be assessed and validated by an independent means. One common approach was to look at the same object in the traditional visible-light microscope at magnifications at which they could be compared. If they looked the same, one could assume that the electron micrograph provided an accurate depiction of the object—that is, as long as one assumed that both images could not be wrong. If the electron microscopic image appeared significantly different, it was presumed to be in error.

Muscle was a particularly good sample material for making such comparisons. The muscles studied by Szent-Györgyi, Heilbrunn, their forebearers, and now the electron microscopists were primarily those associated with locomotion. They are commonly referred to as *voluntary muscles,* because animals voluntarily (or volitionally) contract them; or *skeletal muscles,* because they are attached to bone; or *striated muscles,* because in the light microscope they display characteristic stripes of varying thickness and spacing. It was this third characteristic that was useful to the electron microscopists. One could not only determine whether the various stripes and the spacings between them seen in the light microscope were also seen in electron micrographs, one could assess whether their appearance was *quantitatively* identical. In particular, was the spacing between stripes the same or did it differ, and if so, by how much?

Although the presence of striations had been known for many years, experiments that attempted to discern their nature and role in the contractile process, if any, had in great part failed. Though their meaning had been elusive, still they provided the only substantive clue, along with Hill's energy measurements, about the mechanism of contraction prior to Szent-Györgyi's and Heilbrunn's experiments. One aspect of the pattern of striations seemed to be particularly important. The space between two particular repeating stripes *(Z lines)* decreased when the muscle contracted, seemingly as the result of a shortening of an intervening region with a relatively low refractive index (called the *isotropic* or *I band*). Whatever the mechanism was, it seemed that this shortening had something to do with it.

A Real-Life Parable

If one chose the protocol for sample preparation carefully, the same stripes and spacing, as to number and distances, could be seen in electron micrographs as in the light microscope. When such seemingly accurate images were obtained, Szent-Györgyi's precipitates were nowhere to be found. Nor was there any evidence of Heilbrunn's sol/gel states, though the insides of muscle cells were chock full of interesting structures. Most significantly, there were many very thin fibers that ran parallel to the cell's long axis. The band with the low refractive index (I band) appeared to be an area where the filaments did not greatly overlap. And significantly, when the overlap was greater, the Z lines were closer together.

When contracted muscles were examined, it could be seen that the narrowing of the I band seen in the visible-light microscope was a consequence of more and more fibers overlapping. It seemed that the electron microscope had revealed the cause of this heretofore inexplicable change. The filaments apparently slid past each other as the muscle contracted. From these and a variety of other observations that detailed the structure and organization of the filaments, Hugh Huxley, working in Great Britain, developed a new model for the mechanism of muscle contraction. He called his theory the *sliding filament* model for obvious reasons, and it became the foundation for the modern view of muscle contraction.

Huxley identified two types of filament, one thicker than the other. Not surprisingly, they became known as the *thick* and *thin filaments.* They seemed to be attached to each other intermittently by another structure that Huxley called *cross-bridges,* also for reasons that should be obvious. He proposed that movement of the cross-bridges forced the fibers past each other in a ratchetlike fashion, and that this shortened the muscle. It was the sliding of filaments that accounted for the development of tension in muscle.

Lessons from the Living Cell

The fact that there were two different types of filament and that there were two major proteins in muscle was not lost upon Huxley and other investigators at the time. Indeed, given the fact that muscle cells were filled with these filaments, and that the proteins that Szent-Györgyi had studied were major proteins in the muscle cell, it seemed quite likely that the thin and thick filaments were none other than actin and myosin. Eventually this was demonstrated—actin was the thin filament and myosin the thick one.

Though Szent-Györgyi's mechanism of superprecipitation seemed incorrect, except as a metaphor for the interaction between the two proteins, he had established the two major proteins involved in the contraction of muscle. So, he turned out to be both right (in the identification of the involved proteins) and wrong (about the mechanism). How about Heilbrunn? There was no indication of the state change he had proposed. However, it turned out that the calcium ion plays a central role in the contractile process.

Several years after his initial studies, Huxley attempted to understand what happens at the cross-bridges that causes muscles to contract. A postdoctoral fellow in his laboratory, Saul Winograd, explored the possibility that the calcium ion was involved. To determine this another method was needed—*electron microscopic autoradiography*. Radioactive calcium was added to the medium bathing muscle tissue and given time to enter the cells. The tissue was then prepared for electron microscopy in the manner outlined previously. Finally, in the autoradiographic step, a photographic emulsion was closely apposed to thin sections of the tissue. The location of calcium in the cell could be determined by the identification of small features on the exposed film that looked like spots or squiggles. These features, called *grains*, were regions of the emulsion that had been exposed to isotopic emissions from radioactive

A Real-Life Parable

calcium in the part of the cell that overlay the particular area of the film. We will talk more about this method in Chap. 10.

When autoradiographs of flaccid muscle were examined, calcium was virtually absent from the cross-bridge region. But when the muscle was stimulated to contract, it appeared, apparently having diffused there from elsewhere. It turned out that calcium was stored in a special membrane-enclosed structure within the muscle cell called the *sarcoplasmic reticulum*. The sarcoplasmic reticulum was situated directly beneath the cell membrane—that is, in Heilbrunn's cortex. When the muscle contracts, calcium is released from this depot and diffuses to the center or medullary region of the cell where the cross-bridges are located. This was all just as Heilbrunn had proposed. Like Szent-Györgyi, Heilbrunn was both right (in the identification of calcium's role) and wrong (about the mechanism).

Back to the Beginning

As a result of these and many other studies by a variety of investigators, another aspect of muscle contraction became clear. When the motor nerve to a muscle is stimulated electrically, or *depolarized*, it leads to a similar depolarization of the muscle cell membrane. This in turn causes calcium to be released from the sarcoplasmic reticulum. Calcium then diffuses to the cross-bridge region, where it is bound to a specific calcium-binding site on another protein that, in turn, activates the contractile process. ATP is expended to energize the mechanism, and the cross-bridges, in ways still not fully understood, move and do work in so doing. This in turn causes the fibers themselves to move.

This brings us all the way back to Galen and the notion of extrinsic and intrinsic causes. Szent-Györgyi had uncovered the central chemical substances involved in the contraction of muscle—that is, its intrinsic cause. Heilbrunn had discovered its extrinsic cause, the way in which the process is activated—how the stimulus (a neuronal stimulus) leads to the response (muscle contraction). In so doing, he also illuminated the nature of cellular responsiveness more generally. In the end, Szent-Györgyi and Heilbrunn had been studying different, though intimately related, aspects of the same mechanism.

Often during the active phase of disputes such as theirs, a dialectic takes form. It is assumed that one idea must be right and the other wrong. And while dialectics are very important in science—Hegel argued that only in their presence does the necessary friction exist for progress—the choice between two seemingly inimical views is sometimes more illusory than real. This turned out to be the case in the search for the mechanism of muscle contraction.

CHAPTER 8

The Lesson

> *As one discovers a new phenomenon in nature, one first asks, "What is the phenomenon?"* ... *Once this "what" question is answered, physicists ask, "how does it work?"*
>
> David Gross and Edward Whitten,
> Institute for Advanced Study, Princeton

Beyond the uncertainty of dialectics, what lesson can we take from this story? Let us imagine for a moment that there had been no Heilbrunn, no Huxley, and no electron microscope; and that Szent-Györgyi's model, along with his strong microreductionist inclinations, had continued to dominate thinking to the present day. Let us say further that methods had subsequently been developed that enabled a far more detailed physical and chemical analysis of proteins such as actin and myosin and led to a complete understanding of all of the molecules involved in the contraction of muscle, in each and every atomic and molecular detail, in regard to each and every one of their interactions. But equally, let us also imagine that no other avenues of investigation (neither physiological nor microscopic) were available. The question is whether scientists restricted to such "molecular" circumstances can authoritatively determine the mechanism of muscle contraction from their investigations. According to the strong microreductionist, as exemplified by Szent-Györgyi, this should not only be possible, it is the only way to gain such an understanding.

But whatever the strong microreductionist might believe, the fact is that all that could be learned by exploring the mechanism of muscle contraction in this way is how superprecipitation occurs—*its* mechanism. Adherents to the strong-microreductionist viewpoint would feel that this was tantamount to having unearthed the mechanism of muscle contraction itself. But they would be wrong. They would not be able, at least legitimately, to make inferences about the actual mechanism of muscle contraction from their knowledge of superprecipitation, however sophisticated that knowledge might have become.

We can understand why by asking a question that a skeptic like Heilbrunn might ask. How can you exclude the possibility that what you have discovered is simply an artifact of the in vitro system, and totally unrelated to how muscles contract in situ? Yes, you have shown that the various molecules interact and how, but on what basis can you conclude that this is what occurs in the living cell and produces the contraction of muscle?

Without knowledge of the intact system, the physiological relevance of the precipitation model, or any other similar model, can only be conjecture—a matter for *speculation*, as Heilbrunn said. In its absence, even a theory supported by much molecular detail would be incapable of validation. We could not even be sure that the molecules about which we had learned so much were involved in the process at all, whatever suspicions we might harbor and however reasonable they might seem. Such a proposal can be verified only by consulting the intact system.

The Lesson

On reading this, Epistemon would be quick to disagree. If the scientists' model was incorrect or insufficient, they would naturally discard it, and replace it with a better, more successful one. Of course, it too might be incorrect. But this is simply the normal way of science. Each unsatisfactory model would be discarded and replaced by a more promising one, until, inexorably, the scientists homed in, with increasing accuracy, on the truth of the matter. Examination of the intact system would not be necessary; refinement of the molecular model is all that is needed.

For example, let us say that we could examine the precipitate in the electron microscope and found that it was made of protein fibers. And let us imagine further that we discovered the presence of two types of fiber and their connection via cross-bridges. That is to say, what if the sliding-filament model had been invented through study of the precipitate and its constituent molecules? Would we not then have a successful model obtained by molecular analysis?

But this would change nothing. We still would not know if our model was successful without examining the functioning cell. Bear in mind that we are not simply seeking a model that *might* be explanatory. We are seeking the *one* that *is*. And to establish this, must know how the postulated mechanism plays out in the real space and real time of the intact contractile process.

Nor do we have any right to argue, as is sometimes done, that confirmation in the intact system is simply a prudent control for testing an established molecular theory. That would miss the point. Certainly it is good practice to check results against a control; but crucially, because the parts do not entail the whole, the adequacy of a microreductionist model can be assessed only by considering its goodness of fit to all the observable properties of that whole.

If we had come to the sliding-filament model by exploring the molecular relationships between actin and myosin as isolated chemicals, without any sense of the properties of the whole muscle, the hypothesis would not be verifiable. Indeed, as it was actually invented, the sliding-filament model could not stand on its own. It required validation against the known properties of the intact system. Without that validation, however sophisticated our knowledge, the only way that we can assert a model's correctness is by authority.

The Parts and the Whole

The strong microreductionist sometimes confuses the need to examine the whole living system with what is sometimes referred to as naïve inductivism. *Naïve inductivism* is the belief that phenomena (or properties) can be examined in the absence of theory or hypothesis; that is, they can be viewed naïvely as mere observation without any conceptual baggage. This proposition is usually thought to be false because humans, scientists not excepted, inevitably bring preconceptions of one sort or another to the observation of phenomena. But this is irrelevant here. The scientists are not being asked to examine the properties of the intact system in a naïve fashion at all. They can have all the hypotheses and models they wish. What they are being asked to do is make observations on the natural process that are not *contingent* upon the correctness of any particular model. That is, observations of the properties and qualities of the process of interest are needed that are independent of human hypotheses and theories about the underlying mechanism.

Heilbrunn's idea of a sol/gel transformation was based on an outdated and vague idea about the chemistry of the cytoplasm—the notion of the biological colloid—and it was that seemingly incorrect view that motivated his studies. But nevertheless the observations that he and Wiercinski made were, in and of themselves, descriptive of the process of contraction in the intact muscle cell. For this reason, their observations demanded explanation by *any* successful theory of mechanism.

The same cannot be said of Szent-Györgyi's findings. If his theory of superprecipitation and his conclusion that actin and myosin were the contractile proteins were found to be wrong, his observations would have taught us little, if anything, about muscle contraction. In and of themselves—that is, without reference to the intact process—phenomena discovered by the examination of isolated parts of biological systems can have validity, and perhaps utility, only if the hypothesis that generates them turns out to be true. That is, since such observations *are* contingent upon the correctness of the model, not independent of it, they have no status beyond that model, other than as descriptions of isolated entities and events of unknown physiological meaning.

Even actin and myosin, the two proteins that we know today are central to the mechanism of muscle contraction, are known to serve this purpose only because work such as Huxley's provided a context in which to explain their role within the intact muscle cell. Yes, they were likely choices because of their prominence in homogenates and because of Szent-Györgyi's and others' studies of their properties, but they could never have been more than that on the basis of such information alone.

Process and Mechanism

Thus, processes have properties, and these properties are independent of any particular mechanistic model that we might devise to explain them. And the scientist must not forget that it is these properties that must be accounted for by any model that is claimed to be explanatory. I believe that some of the confusion about this issue concerns conflating the terms *process* and *mechanism*. We have been talking about the need to establish the properties of a *process* to provide the necessary means to evaluate a particular exegetic theory of *mechanism* for that process.

In this sense, a process can be defined as something that changes or evolves over time. It is often thought of in terms of the process of a machine, though processes are not necessarily linked to machines. A machine is designed to carry out a particular process—for example, a series of steps in the manufacture of a product. This process is distinct from the nature and workings of the machine's underlying mechanical parts. Similarly, biological processes are distinct from the underlying chemical and physical means used to carry them out. We can fully characterize and comprehend the function that a machine carries out (the process) without knowledge of the mechanism that underlies that process, but, importantly, not vice-versa.

A good example of this in biology is the transmission of the electrical impulse along the nerve cell or neuron. By characterizing ion and electrical gradients across the membrane of the neuronal axon (the long fine extension of the nerve cell that propagates the electrical impulse), the physical processes that produce electrical conduction along this structure were discovered. It was possible to completely and quantitatively explain its causes, the depolarization and subsequent repolarization of the axonal membrane due to various ions moving across the membrane, *without* any knowledge of or even reference to the detailed biochemical mechanisms that underlie these events. That is, a physiological *process*, the conduction of the nerve impulse in this case, can be fully

The Lesson

understood from studies on whole functioning objects, even though the underlying biochemical *mechanism* is unknown.

But the converse, the guiding principle of strong microreductionism, can never be true. We can never fully infer process from knowledge of what appears to be the underlying mechanism. Consider the neuron. Even if we had complete knowledge of the ion channels, pumps, and other chemical incarnations in the nerve's membrane that underlie the depolarizing and repolarizing events necessary for conduction (we have quite a bit of such knowledge today), we could not infer even the existence, much less the mechanism, of neuronal conduction. We could only say that a particular ion channel was capable of carrying this or that ionic carrier of charge at this or that rate. To connect the movement of ions across membranes to conduction, we have to know in all necessary detail the properties of that conduction itself in the intact cell. Without it, we would lack crucial information, such as the magnitude of the various fluxes, their relationships to each other, and, most important, how everything is connected in time and space to produce conduction.

Perhaps an even more illustrative example is the contraction of heart muscle. Heart muscle is also a striated muscle and functions in a very similar fashion to skeletal muscle. Of course, the heart is the pump that is responsible for the circulation of blood through the arteries, capillaries, and veins that are found in all the organs and tissues of our bodies. We knew with great precision how this muscle "worked"—if by *worked* we mean carried out the process of pumping blood—long before we had any secure knowledge of the underlying contractile mechanism. We were able to characterize the quantitative relationship between the amount of blood entering the chambers of the heart and that pumped out, as well as the relationships between the pressures and volumes involved.

We could even describe in detailed quantitative terms how the length of the heart muscle is related to its ability to pump blood. We knew all of this with mathematical precision in the complete absence of knowledge of the underlying mechanism of muscle contraction. Conversely, if we knew with all possible molecular and chemical detail the mechanism of contraction of heart muscle—say, regarding the sliding of filaments—*from that information alone* it would be quite impossible even to figure out that the heart pumps blood, much less how it does it.

A Final Comment on the Contraction of Muscle

As noted, even though Huxley's experiments with the electron microscope provided crucial information about the intact muscle cell, they were not in themselves sufficient to validate his theory of muscle contraction. The task of correlating the sliding-filament model to the known characteristics of the intact contractile process still confronted the model. Deductive predictions of the theory were formulated and tests attempted. For example, if the theory was correct, then it had to be shown that the extent of overlap and the number of cross-bridge connections seen in the microscope were correlated to the known tension of muscle for a given degree of shortening, or that the time course of contraction and the time course of ATP utilization and cross-bridge phenomena were correlated quantitatively. Some correlations of this sort fit the sliding-filament model relatively well, and although like any model it would continue to be challenged by newly discovered properties of the intact system, it became the standard model. For this achievement a second Nobel prize was awarded for the study of muscle contraction—not to Albert Szent-Györgyi, but to Hugh Huxley.

The Lesson
Second Thoughts

Lewis Heilbrunn had been a teacher of mine when I was an undergraduate at the University of Pennsylvania and was one of two professors who introduced me to biological research. I felt that his contributions, especially in regard to our understanding of the role of the calcium ion in biological processes, had been and is still inappropriately ignored. He not only proposed that calcium was involved in the contraction of muscle and locomotion, as we have discussed; he also proposed, more generally, that it played a central role in the life of the cell. He was ostracized and demeaned by many of his contemporaries for this belief, and considered a person obsessed with what to most seemed a clearly erroneous notion.

Unfortunately, it was not until his death in the early 1960s that attitudes began to change. Obsession or not, he was correct. Calcium serves as an initiator and modulator of many cellular processes. It is especially important as an intermediary when particular stimulants or "irritants" lead to a subsequent functional response from the cell, as in muscle contraction, but also, for example, in the transmission of impulses across nerve synapses and the fertilization of eggs by sperm. In the jargon of the field this is called *stimulus-response coupling,* and it is a central process of life that connects cellular phenomena to the features of its environment.

For his part, Szent-Györgyi remained an active scientist at Woods Hole into his nineties. Some years ago, my wife and I traveled to Woods Hole to collect information for a possible biography of Heilbrunn. We found that Heilbrunn's connection to the institution was still palpable. Indeed, the director had been one of his graduate students, and a surprisingly large number of his students, now mostly retired, summered there—as did his daughter Constance, in the house on the pond. We had the great pleasure of talking to them all—including Floyd Wiercinski, the student who carried out the crucial experiment—about their recollections and memories of the "boss," as they called Heilbrunn.

To my surprise, I was informed that Szent-Györgyi was still alive and at Woods Hole. Not only that, he was apparently busy doing experiments. I decided to approach him to get his opinion about the controversy with Heilbrunn. When I visited his laboratory, I found out that he was very ill, in fact on his deathbed. Though I did not want to disturb him or his family under the circumstances, I realized that this would be my last opportunity to ask about Heilbrunn and the controversy. One of Szent-Györgyi's associates graciously offered to ask the question if an appropriate opportunity presented itself.

On the day before we left, I received a response and was surprised to learn that Szent-Györgyi was pleased to answer my question. In his later years he found himself studying the properties of whole cells in the context of his unsuccessful attempts to understand the nature of carcinogenesis (cancer). In light of his experiences and frustration with this work, his point of view had changed dramatically. He had asked his associate to convey his view that Heilbrunn had been right all along, not about his theory of muscle contraction, but about the study of cells. He had come to realize that one could not learn about the properties of cells solely from a study of their parts. The whole embodied entailments of a sort and nature for which his substantial knowledge of protein biochemistry had not adequately prepared him.

CHAPTER 9

The Making of a Paradigm

> *Normal science, the activity in which most scientists inevitably spend almost all their time, is predicated on the assumption that the scientific community knows what the world is like.*
>
> Thomas Kuhn

As should be clear by now, the need to test mechanistic models against the properties of the intact processes they are thought to account for is a consequence of the fact that the parts of the system do not entail the whole. If they did, then such tests would be unnecessary, and we could fully understand the mechanisms underlying natural processes from microreductionist investigation alone.

We will explore the impact of this conclusion further in this and the following chapters as we consider the second real-life example of the problem with the strong-microreductionist principle—an important contemporary archetype of cellular biology known as the *vesicle theory* or the *vesicle model*. The vesicle theory proposes a mechanism for the movement of various cell products, especially proteins but also other physiologically important substances such as neurotransmitters, both within the cell and to the outside. It is widely believed to provide an accurate, if still incomplete, description of how these processes occur. Like Szent-Györgyi's model for muscle contraction, the vesicle theory was developed in great part from the study of the presumed parts of a mechanism. However in this case, the parts were particular anatomical structures in cells, not chemicals like actin and myosin.

The vesicle theory was arguably the first scientific paradigm—that peculiar mixture of axioms and assumptions, hypotheses and theories, beliefs and prejudices, methods and evidence—to emerge from the application of the electron microscope to the study of the cell. It served as the foundation of a new discipline that became known as *cell biology*. Although cell biology had been in existence from the time that the living cell was first discovered, and at the time of the vesicle theory's invention was being pursued from many different experimental and theoretical perspectives, the term soon came to be associated with a particular type of cell biology in which the electron microscope played a central role. This new cell biology was characterized by its predominantly anatomical context and its strong-microreductionist leanings, as well as its particular mechanistic embodiments, like the vesicle theory. In addition, and importantly, it was typified by an experimental approach in which certain established procedures and techniques, including but not limited to those involved in electron microscopy, were applied to uncover the cell's secrets. We will discuss two of these in some detail in the next chapter.

Although the vesicle theory has dominated thinking in this field for almost half a century and therefore can properly be called the standard model, its validation has been and in important respects remains uncertain and problematic. As

The Making of a Paradigm

we shall see, the theory presents a valuable cautionary tale of the decisive consequences of how we reason from observation to the confirmation of theory, and back again from theory to new observation. My aim in what follows is threefold. First, to illustrate how the strong-microreductionist perspective suffuses this large, well-accepted body of work, down to the most intimate details of particular experiments. Second, to demonstrate that for all meaningful purposes this limited the logical tools at the investigator's disposal to inductive reasoning. This is a serious shortcoming and can give rise to persistent confusion between what is hypothesis and what is proof. And third, to reveal how the vesicle model has been protected from falsification, and that this is a consequence of its strong-microreductionist foundation. Separately and together, these realizations disclose deep methodological weaknesses in much of the formative evidence for the vesicle model, as we shall see.

Until relatively recently the mechanism proposed in the vesicle theory was thought to be universal. That is, it was thought to account for the movement of all intracellular proteins, secreted proteins such as peptide hormones, small organic molecules such as neurotransmitters, and numerous other molecules—whatever the particular molecule, for whatever purpose, in whatever cell. Perhaps more than any other invention, the vesicle model has established how biologists have come to conceptualize the biological cell. It has served as an exemplar of cellular mechanism that embodies the modern perspective, and is antithetical to the old generalist notion of protoplasm discussed earlier. In protoplasmic theory, a single substance was imbued with the properties of life. In the vesicle theory, the cell is envisioned as the ultimate Cartesian machine—much like those in Rube Goldberg's cartoon contraptions or Jean Tinguely's mechanical sculptures—filled with what to the uninitiated may seem a bewildering array of distinct, reciprocating parts, which, driven by a complex ensemble of chemical interactions and reactions, carry out the cell's various functions.

Even at its inception the vesicle model, and later its adjunctive hypothesis called the *signal hypothesis* (I shall discuss this only briefly), offered a very particular and very detailed explanation for how cells sort newly manufactured protein molecules to one location or another. Though we can think of the model in rather general terms, it offers a strikingly complex explanation for this process. For example, depending on how one breaks things down, the basic model (excluding various alternative transport routes) proposes that eucaryotic (nonbacterial) cells require 15 to 30 distinct or separate biological mechanisms to move protein molecules a few micrometers (μm), about 0.0005 in.

Because the model is so specific in what it proposes, experiments have tended to explore the details of its particular mechanistic embodiments, and results have been interpreted in terms of those embodiments. Consequently, the experiments have explored the model's world, not the natural world. Moreover, since the vesicle model, like Szent-Györgyi's superprecipitation model, was in both its origins and development the product of the belief that we can come to understand the nature of systems in their entirety from the study of their parts, it was refined in the absence of substantial information about the intact system it sought to explain.

Heidenhain

The vesicle model originated in and largely remains the product of the anatomical study of cells that are specially designed to secrete proteins. As already discussed, protein transport

The Making of a Paradigm

and secretion are central processes of life. The transport of the cell's various protein products to their specified locations is necessary to establish their functionality. Cellular life, and hence life more generally, cannot exist without such topologically directed movement.

Though the vesicle theory, like the sliding-filament model, was in great part an outgrowth of the application of the electron microscope to biological material just after World War II, its early roots can be found in the great explosion of knowledge about cells that occurred in the wake of the proposal of cell theory during the nineteenth century. In the mid to later years of that century, a major effort was made to describe and classify the different cell types found in vertebrate tissues and organs, especially those of mammals. One of these tissues, the exocrine pancreas, was examined in some detail by the renowned German histologist and physiologist Rudolf Heidenhain. The *exocrine* portion of the pancreas, unlike the more familiar, consanguineous, and far smaller insulin-secreting *endocrine* portion (about 1 percent of its size), is a ducted gland whose secreted products are carried into the intestines via an arborizing system of tubes or ducts. It is the part of the pancreas that secretes the digestive enzymes, the proteins that bear the major responsibility for the digestion of food.

Heidenhain found that this gland was dominated by a single cell type, called an *acinar* cell, from the Latin for berry or grape. These cells were organized in clusters that formed "grapelike" caps at the proximal end of the duct system, closing it off in multitudinous fine terminals. He also discovered that the acinar cells were filled with hundreds of densely staining spherical particles of about 1 μm in diameter.

Wondering what these objects might have to do with enzyme secretion, Heidenhain carried out a series of experiments that were among the earliest to successfully correlate microscopic structure to function. He stimulated pancreatic secretion in dogs by feeding them a meal or by injecting certain pharmacologic agents. He then collected fluid from the pancreatic duct to measure the effect on secretion. Finally, he sacrificed the animals and removed the pancreas for microscopic examination. What he found led to the proposal of the vesicle theory some 75 years later.

Under certain conditions, the area of the acinar cell occupied by these tiny spherical objects became smaller when secretion was stimulated. Heidenhain recognized that this could be explained in three ways. The objects might have moved closer together; there might have been fewer of them; or they might have become smaller. Of course, all three events could have occurred: fewer objects, having become smaller, might have moved into closer proximity to each other.

He drew two conclusions from this observation. First, he reasoned that the anatomical change he had observed indicated that the secreted enzyme came from these cells and their small, dense ball-like structures. He proposed that it was these tiny particles that stored the pancreatic enzymes and that were the source of their secretion. This would explain why they occupied a smaller proportion of the cell after secretion had occurred. Digestive enzymes were often called *zymogens*, and Heidenhain called these particles *zymogen granules* to denote that they contained digestive enzymes.

Zymogen granules are a particular type of structure that we identify today by the more general term *secretion granule* to specify their role in secretion. Although Heidenhain's inference about zymogen granules was widely accepted and seemed reasonable given his observations, it took another 75 years for it to be proven true. It was only in the 1950s, when they were actually separated from the acinar cells, that

The Making of a Paradigm

zymogen granules could be chemically analyzed and shown to contain digestive enzymes. Today we know that similar secretion granules are found in many different cell types, where they store a wide range of biologically important secreted substances in addition to digestive enzymes, such as peptide hormones and neurotransmitters.

Heidenhain's second conclusion was that the granules were reduced in both number and size. As with the first conclusion, he could not confirm this judgment. In this case it was because of the limited quality of his microscope's optics. In practice, its resolution was slightly less than the size of the objects being examined, and although he could see them well enough to identify their presence, he could not measure their size or number accurately, much less changes in these parameters. It was not until the high-resolution electron microscope became available that such measurements were possible. Then it was determined that Heidenhain (and two contemporaries, Kuhne and Lea) had been correct. Granules can decrease in both number and size during the active secretion of digestive enzymes.

Heidenhain did not consider potential mechanisms for granule shrinkage, but instead focused his attention on how their number might be reduced. He imagined that they were lost by expulsion from the cell in toto, contents and all, as opposed to the alternative—their disappearance *within* the cell. The idea of expulsion made sense to him, since the need seemed to be to secrete the contents of the granules *outside* the cell, not inside it.

Though he did not have any substantive evidence on the matter, the idea that secretion was the result of the ejection of zymogen granules from the acinar cell captured the imagination of Heidenhain's contemporaries as well as the microscopic anatomists of future generations. Numerous variations on the same theme were proposed over the next 75 years, based on observations in the visible-light microscope. On occasion, investigators reported seeing granules actually pop out of cells. But the resolution of the visible-light microscope did not allow for a clear assessment of such an event, any more than it allowed one to measure changes in granule size. As a result, such claims were thought to be either visual illusions or the consequence of some sort of artifact.

The Electron Microscope and the Vesicle Theory

Matters remained very uncertain, not only about zymogen granules and what happens to them, but about the internal structure of the acinar cell. Indeed, the internal structure of cells in general was an extremely uncertain business because of the low resolution of the available microscopes. Often only objects of striking appearance, such as the densely staining zymogen granules, or very large ones such as the nucleus, could be identified with any clarity. Indeed, it was argued by many that there was nothing else there to be seen.

This was the climate when the electron microscope exploded onto the scene. Its application to biology changed our view of cells forever. Not only was it patently mistaken to conclude that there was nothing in cells other than what had been observed with the visible-light microscope, but a remarkable new world of cellular structure, a world of incredible variety and density, was discovered.

As already mentioned, these discoveries put the final nail in the coffin of the protoplasmic theory of life, and simultaneously gave rise to the modern concept of the cell. In a veritable orgy of investigation, methods were developed that exposed the fine structure of just about every known cell type. A new group of cellular biologists—sometimes referred to as *electron microscopists*—was born whose task was to illuminate the internal structure of cells with this high-resolution device.

Cellular content could be seen with a structural clarity that could only have been imagined before. Among the cells examined, pancreatic acinar cells were particularly interesting. Previously it had only been possible to identify the nucleus and zymogen granules with any certainty. Now it could be seen that this cell was filled with structures. Some were seen for the first time, while others that had been suggested by visible-light microscopy could now be confirmed.

Function from Structure

Following Heidenhain, it was natural for electron microscopists to correlate the structures they uncovered to the known functions of cells. On the basis of their interpretative tendencies, electron microscopists could be divided into two broad groups, the conservatives and the progressives.

The *conservatives* drew functional conclusions from their observations with great caution. They believed that as powerful a research tool as the electron microscope was, what they were examining was still far from the living cell. In addition to being sliced thin, purged of water, embedded in plastic, and doped with heavy metals, the contents of the cell had to be fixed in place chemically before one could do anything. An old method also used in visible-light microscopy, fixation was an attempt to link all cellular chemicals and structures to each other to prevent their movement or loss during further preparative procedures.

After all of these transmutations, the sample had to be placed in a vacuum, where it was bombarded by high-energy electrons. How these procedures changed the objects being examined was unclear, and remains a concern today in spite of the revolutionary impact of the method. Finally, and crucially, dynamic observations were out of the question. Certainly, after all of these preparative procedures the samples were no longer alive. And in any event, how could living samples be examined in the vacuum of the electron microscope? For these and a variety of other reasons, the conservatives concluded that electron microscopic images could not *in their own right* connect structure to function. They might bring possibilities to mind, hypotheses to be tested in the whole functioning system. That was a great deal, but that was all they could do.

I call the second group *progressives* because their desire to make progress made them more willing to draw functional conclusions from electron micrographs. The progressives thought that if they described the structures in the cell and their juxtaposition to each other well enough, they could with thought and care infer the character of the functioning whole. As such, they were believers in the strong-microreductionist principle.

The Acinar Cell

This progressive attitude was nowhere more evident than in electron microscopic studies of the structure-laden acinar cell. As much as it was natural to connect the filaments seen in the muscle cell to the contraction of muscle, it was natural to connect the structures in the acinar cell to the secretion of digestive enzymes, though these anatomical features were also found in one form or another in widely varying perfusion in most eucaryotic cells, even muscle cells.

The Making of a Paradigm

In any event, and however common in other cell types, all of the major structures in the acinar cell (with two exceptions, the nucleus and the mitochondrion) were proposed to be involved in digestive enzyme secretion. Some of these structures had been seen with the visible-light microscope, but with the exception of the zymogen granule, their existence, much less their provenance, had been uncertain. In any event, nothing much had been learned about them from light-microscopic studies. One just could not look closely enough to make out their form and detailed organization. At best they were vague apparitions.

Two structures that had been suggested by visible-light microscopic studies, the *Golgi apparatus* (named after its discoverer) and the *endoplasmic reticulum* (ER), turned out to be major features in electron microscopic images of the acinar cell. The ER was particularly striking. It appeared as a series of elongated membranous sacs organized in layers or stacks that occupied some 20 percent of the cell's volume. Although the ER is found in most eukaryotic cells, it was particularly well developed in the acinar cell. Initially it was called the *rough-surfaced endoplasmic reticulum* because, unlike other "smooth" intracellular membranes, its surface was studded with many small (about 15 to 30 nm), seemingly spherical particles that gave it a rough appearance. The same particles were also found in the cytoplasm, apparently free and unattached. It was discovered that these particles contained substantial amounts of RNA and were named *ribosomes* for this reason. Ribosomes turned out to be objects of great importance. It was discovered that protein synthesis (manufacture) took place on them—and beyond DNA replication, this was the central chemical process of life. And because many ribosomes were attached to the membranes of the ER, it was natural to associate the ER itself with protein manufacture.

The acinar cell was polarized. It had an apical pole that faced the duct system of the gland, where the secreted product left to travel into the gastrointestinal tract, and a basal pole that faced the tissue's interstitial or extracellular spaces and its capillary blood supply, which provided the cell with nutrients. The ribosome-containing ER was located at the cell's basal end. Thus, it seemed that new protein synthesized on ribosomes attached to the ER subsequently traveled to the cell's apical pole, where it was released into the duct system. That is, to be secreted new protein had to move from one pole of the cell to the other. Heidenhain's zymogen granules, the storage repositories for the digestive enzymes, were located near the cell's apex, adjacent to its enclosing membrane. The task then seemed to be to understand how new protein moved from its site of synthesis on the ER at the base of the cell to the zymogen granules at its apex and from there into the duct system. It was this perceived need that led to the development of the vesicle theory.

Like the ER, the prominent Golgi apparatus also seemed to be composed of a stack of elongated sacs, though lacking ribosomes, far fewer in number and less regular in appearance than the ER. Because the Golgi was situated neatly between the ER and the zymogen granules, it was proposed to be a way station on the path from ER to zymogen granule. On the edge of the Golgi apparatus, seemingly between it and the ER, were numerous small (roughly 40 nm in diameter) spherical objects that were called *small vesicles* or *microvesicles*. These tiny vesicles were thought to be conveyances that moved new protein from the ER to the Golgi apparatus, among the various elements of the Golgi apparatus, and eventually to the zymogen granule—like so many minibuses shuttling the cell's protein products here and there.

The Making of a Paradigm

Before we go further I should explain what I mean when I call something a *vesicle*. From an anatomical perspective, essentially all membrane-enclosed structures in cells that are not grossly elongated (that is, that are more or less spherical), that have no internal membranes, or that have not otherwise been characterized (such as the nucleus) can be and are often designated as vesicles. Vesicles can vary in volume over some six to eight orders of magnitude. Today, the 1-µm diameter zymogen granule and all other similar secretion granules are classified as vesicles, as are far smaller objects such as the 40-nm-diameter microvesicles just mentioned. However vague this anatomic characterization may seem, the purpose envisioned for vesicles is not vague at all. A vesicle is any membrane-enclosed structure found in a cell that acts as a *vehicle* to transport molecules from one place to another. It is the understanding that vesicles move in a directed fashion, and as a consequence move their contained substances, that is implicit whenever we identify an intracellular structure as a vesicle. As we shall see, though it is easy enough to call an object a vesicle and to imagine it moving, carrying substances, and doing the other things that vesicles are thought to do, such as being formed from and fusing into membranes, showing that this is actually the case turned out to be a difficult task.

Anatomic Fidelity

Functional models based on anatomical descriptions have many problems, as we shall see. To begin with, they can only be as good as the description itself. Among other things, the description that I have just presented is an *idealization*. For example, even though the major portion of the endoplasmic reticulum is found in the basal region of the cell, it quite commonly extends into the cell's apical region. Likewise, while zymogen granules are found predominantly at the apical pole of the cell, they are also found in the basal region, in some circumstances in relatively great profusion. And although the Golgi apparatus is usually situated more or less between the endoplasmic reticulum and zymogen granules, its exact location varies substantially. The same is true for the location and number of small vesicles.

The description I have presented is also an idealization in that it offers conclusions about the anatomical features of complex three-dimensional structures arranged in an intricate three-dimensional juxtaposition to each other from what are essentially two-dimensional images. They are a two-dimensional Flatlander's conclusions about the three-dimensional world. As said earlier, electron micrographs are usually made of very thin sections of the cell and its substructure—on the order of 80 to 100 nm thick. Because the cell's components range in size from roughly 1000 to 15,000 nm in diameter, only some 0.5 to 10 percent of their thickness is usually sampled. Thus, the nature of the whole can only be inferred, not directly apprehended, from such images.

There have been a variety of attempts to do serial sectioning of tissue (cutting series of adjacent sections) for electron microscopic viewing so that the three-dimensional whole can be reconstructed from the superimposition of neighboring sections, but this has not been very successful. To my knowledge, such images have never been reported for the structures we are discussing here. In any event, it is very easy to misconstrue shape, form, arrangement, and juxtaposition when extrapolating two-dimensional sections to the three-dimensional world. It is like trying to reconstruct a car from a slice through 1 to 10 percent of its thickness—not only without prior knowledge of just what the object is, but without any knowledge that such a thing as a car with a particular structure exists.

The Golgi apparatus offers a good example of this problem. It had been imagined from thin sections that the Golgi apparatus was a particular grouping of separate flattened membranous pouches. When it became possible to examine such structures whole or almost whole using surface-scanning techniques, it seemed instead to be a single roundish object with numerous intricate channels and protuberances, not a group of separate flattened sacs. Though making these observations required extensive manipulation of the sample—and the true structure of the Golgi therefore remains a matter of some speculation—it is easy to understand why it would be thought to be composed of separate flattened sacs if all one could see were thin slices through the structure.

If the simple anatomical description I have given is an idealization, it is also an *hypothesis* for structure, and not a true description of it. Whether we are talking of the geometric uncertainties or the fact that objects must be chemically and physically transformed in order to examine them, all standard electron micrographs are hypotheses for, or models of, natural structures, not descriptions of them. There is often good reason to think that the images bear a reasonable relationship to the natural object, even one of substantial fidelity. But as we shall see, however faithful they may be, the hypothetical nature of the images becomes an important problem when the strong-microreductionist perspective makes this uncertainty seem inconsequential or irrelevant.

For example, the various structures in the acinar cell (ER, Golgi, zymogen granules, small vesicles, etc.) are widely believed to serve specific purposes. To do so they must have particular shapes, forms, arrangements, and juxtapositions both within themselves and in relation to each other. That is, for us to draw inferences regarding structure and function—to conclude that these structures serve the particular purposes we infer from experimental work—we must allow that their shapes, forms, arrangements, and juxtapositions are *established,* not merely assumed.

It may seem surprising to those not familiar with this field—and probably not at all surprising to those who are—that there can be great confidence in a functional model when a central aspect of its construction, the presence of particular structural features and the spatial relationships between them, is uncertain and in some respects unknown. But such confidence has abounded for half a century. To the strong microreductionist, if our knowledge of the parts of a system seems sufficient, even parts of parts, we can infer the whole with assurance.

The Genesis of the Vesicle Theory

In addition to the anatomical description of the cell that I have just presented, three other electron microscopic observations were important in the genesis of the vesicle theory. The first concerns the lateral ends of individual flattened saccules of the ER. At their ends the saccules are usually free of ribosomes. They are said to be *denuded* there. Occasionally a broadening or swelling is seen at this site, and this gave observers the impression that they were watching the membrane in the process of budding from the reticulum to form the numerous small vesicles seen between the ER and the Golgi apparatus.

The second observation was that even though most zymogen granules appeared to be uniformly filled with an opaque material (presumably a high concentration of protein)—and indeed these are the most opaque objects seen in electron micrographs of this cell—there was another smaller group of similar vesicle structures mixed among them that were of approximately the same size but appeared to be either empty or partially empty. It was imagined that these were zymogen granules that had yet to be filled. They were called *condensing granules* or *vacuoles,* because it was thought that some sort of condensation of protein occurred within the granule during the filling process.

The third observation was that the surface membrane at the cell's apex occasionally had a shape similar to that of the Greek letter omega. Although structurally quite variable, such areas display a more or less hemicircular center bounded by two more or less flattened arms. This form suggested to investigators that a zymogen granule had just fused with the cell membrane and released its contents. The omega figure was the modern version of Heidenhain's idea of zymogen granules popping out of cells. Instead of popping out, the membrane of the granule fused with the cell's membrane. This in turn produced a hole in both membranes through which the material in the granule could leave the cell and enter the extracellular environment.

From these observations, as well as a few others that I will not go into here, the vesicle model took form. It can be outlined as follows. Proteins are synthesized on ribosomes attached to the ER. The new protein then enters its internal spaces, called the *cisternal spaces,* presumably through a small pore in the membrane underlying each ribosome. How this is supposed to happen is detailed in the signal hypothesis. The signal hypothesis is an intricate hypothesis in its own right, and offers a glimpse into the remarkable complexity of the vesicle theory. This single step in the overall process, the transfer of new protein into the cisternal spaces of the ER, is thought to require 5 to 10 distinct events or interactions.

After entering the ER, proteins destined for transport out of the cell or to other parts of it leave its internal spaces in small vesicles that bud from ribosome-denuded ends of reticular membranes. The small vesicles then migrate to the Golgi apparatus, where they deposit their protein contents after the fusion of the vesicle membrane with that of the Golgi sac. In the modern version of this theory, there are four separate Golgi saccules—cis, medial, trans, and trans-Golgi network—through which the proteins must pass in sequence. This occurs as a result of the formation of new vesicles that contain a particular product or group of products that bud off of each particular Golgi saccule. The new vesicle then moves to the next saccule in line, where it deposits its contents after its membrane fuses with that of the recipient sac, and so on through the four compartments. Having passed through the Golgi apparatus in this way, the protein is packaged again in small vesicles that then move to the condensing vacuoles. The small vesicles fuse with the membranes of these larger structures and release their contents into their internal spaces. After several hundred thousand or so of these events, a condensing vacuole is filled and becomes a zymogen granule. Finally, when a stimulant elicits secretion, the granules move to the apex of the cell, where the last fusion event takes place. This time the secretion granule fuses with the cell membrane, and granule contents are released into the duct system. This last step is known as *exocytosis*, and it gives rise to the omega figure.

This then is the vesicle theory of protein secretion. As complex as this may seem to the newcomer, what I have given is a rather sketchy outline of a far more involved model in terms of what is explicitly proposed, implicitly required,

The Making of a Paradigm

or otherwise implied. Perhaps you can imagine some of the detailed steps that would be required to accomplish what I have just summarized. Each event calls for an underlying mechanism to accomplish the stated aims, such as the budding of vesicles, the partitioning of particular proteins in particular vesicles, the directed movement of the vesicles, their fusion to specific membranes, the development of holes in two fused membranes, the transfer of material directionally between vesicle and recipient compartment, and so forth.

And there can be no confusion—at least no significant confusion—in this transit system, such as vesicles fusing to the wrong membrane. If this happened with any frequency, the vesicle's passengers would of course end up at the wrong location—much like boarding the A train to Harlem and ending up instead at Coney Island. The cell could not survive many such disordering events. As a consequence, each step in the sequence has to be distinguished from every other step, and each vesicle correctly sorted as a function of its contained products. Proteins destined for intracellular locations, such as enzymes targeted to the lysosome (a structure that degrades endogenous and exogenous proteins), cannot end up in the zymogen granule to be secreted, or vice versa for digestive enzymes. In any event, as I have said, the vesicle model calls for dozens of specific events and interactions to move molecules over extremely short distances. It is among the most complex models ever proposed for a biological process and certainly the most elaborate biological transport system ever envisioned.

Thus, primarily from anatomical observations, a complex theory of mechanism was developed for the movement of proteins from their site of synthesis to their point of release from the cell. In its early incarnations, beyond knowledge that protein synthesis occurs on ribosomes and that zymogen granules store the secreted products, the theory was supported by little other than the visual appearance and apparent juxtaposition of objects seen in electron micrographs of the cell, with one significant exception that we will discuss in the next chapter.

Most important, *it was not known whether any of the proposed vesicle transport processes actually occurred.* That is, whether we are referring to the movement of vesicles from the ER to the Golgi apparatus, among the Golgi stacks, or from there to the filling secretion granule, or of movement of the mature granule to the cell membrane, there was no evidence that any of the vesicles moved at all, much less in a directed fashion to a specified site.

And perhaps even more startling, with the exception of the zymogen granule, it was not known whether any of them actually contained the products that were moved. Needless to say, given these facts there could be no quantitative evidence concerning how much material moved in the way proposed. A complex, dynamic model for the transport of various products manufactured by cells had been developed in almost total ignorance of whether the posited events actually occurred. It was on these slim reeds of the imagination that the vesicle model was built.

The Emergence of a Paradigm

Whatever its evidentiary shortcomings, the vesicle theory became popular in short order. It provided a concrete and detailed model for the transport of enzymes, peptide hormones, neurotransmitters, and so forth where none had existed. Perhaps its popularity was only to be expected. The theory assured scientists that an explanation for an important

The Making of a Paradigm

process was in hand, or at least close. But however predictable, its rapid acceptance was also quite extraordinary.

Whether they thought the evidence convincing or weak, or were ignorant of it, many biologists soon came to believe that the vesicle model was not a model at all, but a description of an actual mechanism. The scientific literature reflected this sense of understanding. Papers commonly talked about the model, either in general or regarding particular aspects of it, as known and secure events of nature. Textbooks followed suit by communicating this comprehension to students. Such descriptions did not highlight, or commonly even mention, the immense lacunae of ignorance, the various unanswered questions about the model's fidelity to the natural process. Instead they gave the impression that it was all known, or at least almost all known, a certified product of laboratory research.

The vesicle theory achieved a prominent place among known biological mechanisms during the 20 years following its initial proposal. Remarkably, this occurred without *any* substantial improvement in its evidentiary base. Although much additional research was published on the subject, central, indeed crucial, evidence was still lacking. There were certainly more pictures, of better quality, in different systems and circumstances, but no one had yet found a way to actually observe any of the proposed events in living cells, either with a microscope or in other ways. Neither the budding, fusion, or movement of any of the vesicles thought to be involved in the system had been demonstrated. In spite of 20 years of work, with the exception of one important line of investigation, evidence for the vesicle theory remained essentially static both literally and figuratively. However widely held, it was still an unproven theory.

It is important to be clear that the vesicle theory was not simply a favored and popular theory that was nonetheless understood to have substantial evidentiary shortcomings. The specific mechanisms proposed were widely thought to account for the transport of proteins and other relevant substances wherever and whenever it occurred. And this belief went beyond the mere conviction that such a mechanism existed. It was commonly believed to be the sole mechanism. And this, quite remarkably, in the complete absence of quantitative data. Some had doubts, but with few exceptions kept their own counsel. When questions were raised publicly, they tended to be centered around specific uncertainties, questions of detail, in a known mechanism. The vesicle model was not thought to be a model at all, but an event of nature.

When evidence was reported that did not seem to fit the theory, it, not the theory, became suspect. This or that particular discomforting result could be assumed to be an artifact, incorrectly obtained, or otherwise inadequate or insufficient. For example, the absence of omega figures (the anatomical forms thought to indicate the prior occurrence of exocytosis) or other anatomic forms predicted by the model where and when they were to be expected did not raise doubts about the theory, but instead led to an expanded search for the missing form.

Sometimes negative observations were, as it were, turned on their collective heads, and viewed as evidence in support of the theory. Or, even more remarkably, were seen as providing new detail about the workings of the vesicle mechanism. For example, if omega figures were hard to find, it was because the events involved in exocytosis occurred too rapidly for them to be seen. That is, their scarcity was proof of the rapidity of a known mechanism. Or if the frequency of appearance of omega figures was not increased when the rate

The Making of a Paradigm

of secretion was increased, as would be expected however rapid the event, then stimulation of secretion must have increased the speed at which exocytosis occurred, and that was why no increase in their numbers was seen. Or if secretion vesicles were not found in a particular cell that nonetheless was known to secrete a given product, then it was understood that although they were present they could not be observed, because there were only a few of them, and they filled, moved, and discharged their contents very rapidly—not that they did not exist there and that secretion occurred by other means. Often such knowledge was thought to demonstrate the robustness and essential correctness of the theory, because it provided additional detail about the mechanism.

And when negative observations did produce modifications in the model, its essential features were invariably retained. They did not lead to its rejection, or even weaken its hold, but instead gave rise to its elaboration. If a particular vesicle could not account for what was observed, then an additional type was posited—for example, a vesicle that carried different substances or moved them along a different path. Such new vesicles were not proposed because of substantive independent proof of their existence, but because the seemingly inconsistent phenomena seemed to require them. Unseen or unknown, in an act of circular reasoning, the need was father to the fact.

And finally, when negative observations were thought inexplicable and could not be conveniently rationalized, they were placed in the bottom drawer to await future knowledge that would, it was confidently assumed, explain them within the borders of the vesicle theory. In these and various other ways, negative observations were easily accommodated to the theory, either leaving the model wholly intact or modifying it only in its details. Almost anything that was observed, it seemed, could in one fashion or another be fit into the theory. And wasn't this an indication of its cogency, indeed its correctness?

As a consequence of this mind-set, a potentially useful model, whatever its deficiencies, became a dogmatic belief whose overarching schema was never challenged. Significantly, with exceptions we shall discuss in Chaps. 11 and 12, during all these years the theory was never seriously tested against the properties of the intact functioning system, either in its own right or in comparison to alternative conceptualizations, much less in a skeptical manner.

It seemed that the vesicle theory was protected from falsification—immunized, invulnerable, whatever observations might be made. Evidence of all sorts, affirmative or negative, invariably, and seemingly inevitably, was interpreted to bolster the theory's claims. In this way the vesicle theory was *explicated* by experimentation, but never *threatened* by it. But a theory that is not in jeopardy of falsification is not a scientific theory. It can only be a set of *untested* claims regardless of how widely held they might be or how much evidence can be brought to bear in their support. And a theory that is thought true but untested is indeed nothing more nor less than dogma.

Induction and Deduction

This sorry state of affairs can be attributed in great part to the strong-microreductionist principle. The interpretations that were made were based on inferences from what were thought to be parts of the system to an unknown functioning whole. Because that whole and its properties were unknown, who could say that the observations did not indicate the presence of the mechanisms posited by the vesicle theory? Might not

The Making of a Paradigm

the omega figure indicate the occurrence of exocytosis? Could not the occasional dilations seen at denuded ends of the ER indicate budding, and so forth? Of course they could; who could say otherwise?

And certainly, asking such questions was both natural and perfectly appropriate. But all too often, and with little or no additional information, such unanswered questions became declarative statements. The omega figure *revealed* exocytosis, and the swelling of denuded ends of the ER *disclosed* budding. This was the vernacular of the field—in paper after paper, static observations were infused with evocative dynamic language and presented as proof of the correctness of the vesicle theory. It seemed to many that nature's mechanisms had been exposed by such strong-microreductionist inferences.

The reasoning was inductive. In induction one seeks to infer nature's laws and mechanisms from observation. An observation, say an omega figure, is made and the investigator imbues it with a particular meaning, infers its meaning inductively, say the occurrence of exocytosis. This is what I see, and this is what I believe it means. In this way, a sense of the whole can be inferred from an examination of presumed parts of the system.

There is nothing wrong with this in and of itself. Indeed, such inferences are a central tool of science, most certainly of biology, and are probably the primary means used to generate hypotheses and theories. But this is the crux of the matter. *Induction generates hypotheses and theories. It does not prove them true.* It was this confusion that took place as the vesicle theory gained currency. Inductively spawned hypotheses came to be seen not as hypothetical at all, but as proven aspects of nature. In an act of circular reasoning, an observation both generated an hypothesis and simultaneously served as proof of its correctness.

To make this explicit, let us use the omega figure as an example. At some time a scientist seeing a configuration of this sort realized that such a geometric form might be seen subsequent to exocytosis. This is an hypothesis, an inductive inference based on the evidence of the omega figure. This is what the investigator saw, and this is what the investigator believes it might mean. But the hypothetical nature of the inference was either quickly forgotten or just not appreciated. The omega figure came to be viewed as experimental evidence for exocytosis. That is, when such shapes were seen on the surface of cells, they were considered proof of prior exocytosis at that site. This way of thinking was memorialized when the term *omega figure* was replaced by the mechanistically explicit term *exocytosis figure,* implying an evidentiary connection between the omega figure and a particular event—exocytosis.

The problem was that no such connection existed. There was no proof to be interpolated between the hypothesis that the omega figure might reflect such an occurrence, and the conclusion that it did. They were one and the same. How could one know that the omega figure did not signify something else? Indeed, perhaps it had no functional meaning whatsoever. Maybe it was just a random configuration of the cell's surface that is seen from time to time. Or perhaps it indicated the prior fusion of granule and cell membrane all right, but as an artifact, not a physiological process, say due to fixation. In the end, who could say what this shape meant, if all we knew was that it existed?

This problem in which an hypothesis is confused with proof, theory with understanding, is an example of the philosophical problem of induction. It is a difficulty that is deeply embedded in the strong-microreductionist way of thinking. Even though observations customarily generate hypotheses

The Making of a Paradigm

by inductive reasoning, when we restrict our practice of learning to inductive inferences, such observations often come to be seen as proof for the hypotheses they generate.

To validate an inductively generated theory we must apply deduction. We must ask whether the properties of the natural process are consistent with those predicted by our theory. To the extent that deductive inferences can sanction confidence in the correctness of a theory's claims about nature's mechanisms, it is by testing that theory against the natural process. We formulate deductive hypotheses with the following structure: such and such should happen if our theory is correct; we then seek to validate that particular prediction. What are deduced are usually subsidiary hypotheses of a broader, more encompassing general theory, and it is these hypotheses that we then test against nature. If the properties predicted by the theory are not found in the natural world, then our theory is said to be incorrect or false. If a series of deductive predictions is borne out by the evidence, then the theory is validated, unless and until it fails other deductive tests.

Even though the central and critical utility of deductive hypotheses is as a means of testing our theories against the realities of nature, the strong microreductionist rejects this operation by choice. This is because the strong microreductionist judges exploration of the intact natural system as being unnecessary, and believes that full understanding can be achieved in its absence from the bottom up. The strong microreductionist argues, usually implicitly (by the nature of the research itself), that it is not necessary to test scientific models against the natural system *as long as one learns well about its presumed components.* As said, to the strong microreductionist, the truth or falsity of a functional model can be fully assessed by the in-depth study of parts.

When this is attempted, scrutiny of a theory's validity is determined wholly within its confines, within its own borders. And while deductive logic can still be applied, it is anemic and involutional. Let us use the omega figure again to show how this works. Having inductively generated the hypothesis that the omega figure attests to the presence of exocytosis in a particular cell, the strong microreductionist might seek to test this hypothesis by asking whether similar forms are found in other cells that are known to secrete particular substances or that contain secretion granules, cells that have yet to be examined. The deductive hypothesis takes the following form. If the omega figure is an indication of a prior exocytic event, then we should find these forms in other cells known to secrete one substance or another if exocytosis is the mechanism that underlies the process. Because the omega figure is an object of nature, we are in a sense performing a deductive test of our hypothesis against nature. If these forms are found in these other systems, it provides additional proof of the hypothesis' correctness.

But this is just not true. All the investigator is doing is recycling his or her hypothesis in the guise of proof. The investigator is simply applying the inductive hypothesis about the omega figure to another cell. The fact is that the new observations are no different from the old, original ones. All they allow is the generation of the same hypothesis for new cells; an inductive inference about omega figures for them. Nonetheless, whatever the logical deficiencies of such proof, it has often been considered additional evidence of the theory's correctness. Not only is the exocytosis mechanism found here, but there as well. But of course the meaning of the same evidence is no less hypothetical for having been seen in various circumstances. It is something like concluding that we have been visited by extraterrestrials because so many have reported flying saucers.

The Making of a Paradigm

Everything invariably collapses back into the model itself. The omega figure is not seen as a geometric form that might have a variety of possible meanings. It is conceptualized *only* in terms of the exocytosis model. And when additional proof is sought, it is sought in this context and only in this context. If we adhere to this line of reasoning, we can be confident that we are learning about nature in a significant way only *if* our model turns out to be correct, not regardless of whether it is correct. Even though all observations of nature, however restricted, seen through whatever lens, are de facto expressions of nature's properties, if we do not consult the intact natural system—whether atom, molecule, cell, or whole organism—we have no guarantee that the connection between our observations and the mechanism envisioned by our model is one and the same as nature's, or even one of substance in regard to nature's. Consequently, much of what the strong microreductionist learns may simply be embroidery of the investigator's own model, even though on a cursory examination it may appear to be an expression of nature.

The central point is that models and theories arrived at inductively represent merely the beginning of the search for understanding, not its culmination. Deductively constructed hypotheses serve the purpose of assessing the correctness of inductively generated models and theories against the properties of nature. And observations generated by such deductive hypotheses have assured validity *regardless* of the soundness of the model that generated them only if they are made on the intact system and describe its particular properties. Only then do observations have a meaningful existence independent of the theories of mechanism that give rise to them. In the absence of an exploration of the properties of the intact process, we see nature only through the myopic eyes of our theories, never confronting those theories critically with an unimpressed nature.

Back to the Vesicle Model

As I have said, the core principle of the vesicle theory—that the transfer of product occurs by means of vesicle formation, movement, and fusion or fission both within and to the outside of the cell—was without direct experimental support when it was first generated. This in itself is not surprising. Scientific investigation often begins with models and hypotheses that are little more than the product of an investigator's fertile imagination, the aforementioned inductive inferences. And it is traditional in the microscopic analysis of biological material to imagine dynamical events when they cannot be directly observed. It is what happens after such models and hypotheses are invented that is critical.

We should be clear that whether a researcher decides to test a hypothesis against nature is not a matter of choice. Science requires it. Yet in regard to the vesicle theory such tests were neither carried out nor thought necessary, nor are they for any strong-microreductionist construction. Instead, attention is focused on establishing the details of the particular events proposed by the model.

This approach was justified for the vesicle theory in two ways. First, as we have been discussing, in accordance with strong-microreductionist thought, it was believed that if one learned enough about the parts of the putative system, that knowledge alone would be enabling. Second, there was great confidence that the vesicle model was correct in a general sense a priori; that is, the overall outline was thought to be correct, and necessarily so, even in the absence of evidence. This view was grounded in various assumptions about the properties of the secreted products and the cell that were thought to be axiomatic, and that if accepted left one with the conclusion that it was the only possible mechanism. In the fullness of time this has been shown to be incorrect, and we will discuss the most important of these assumptions in Chap. 12. But be that as it may, if the assumptions were truly justified, then there was no reason to look beyond the vesicle model. As a result of these two attitudes, there came to be substantial confusion about what was really known about the transport of material through and out of the cell and what was only supposed. There was much confusion between what was proof and what was hypothesis.

After the Paradigm

This was particularly so during the 20 or so years following the initial proposal of the vesicle theory, and in many respects remains so today. Observations that seemed consistent with such a mechanism overwhelmed the fact that the theory had never been properly tested against the natural system. At meeting after meeting, symposium after symposium, its "correctness" was affirmed. In this light, given the theory's importance in the description of the modern cell, it came as little surprise when the Nobel prize was awarded to certain of its inventors in 1974. With this important imprimatur, the character of the Cartesian cell, filled with vesicles moving to and fro, in and out, had seemingly been established, even though in point of scientific fact there was no substantive proof (with one exception) for the existence of such mechanisms.

Nonetheless, a new scientific paradigm had been established that was applied not only to the particular system that had been studied intensively initially, the pancreatic acinar cell, but to many other cellular processes, including the important mechanism for the transmission of the electrical impulse across the nerve synapse. A complex system of understanding had been fixed in place that was part theory, part evidence, part method, part axiom, and part unsubstantiated belief.

Though many who consider the vesicle theory true will no doubt dispute this conclusion, some might be willing to grant some of what I have said, at least in retrospect. They might admit that we knew far less in the past than today, and that perhaps the model had not been fully proven in 1974. But this, they might argue, has turned out to be irrelevant. Whatever was the case in the past, *today* it has been established. Also, I am being too philosophical, they might say, and not sufficiently in touch with the real world of practical research. Science is not as neat and logical as philosophers imagine. It proceeds by intelligent guesses, sheer luck, and simple hard work. Yes, reason and logic are important, but we cannot be too rigid in insisting that they be followed to the exclusion of the best guesses and common sense of our most intelligent practitioners. The guesses that were made in the past were reasonable guesses and, as it has turned out, good ones.

But one could board a time machine and find the same confident statement of proof being made soon after the vesicle model's original proposal, and certainly 20 years later when the Nobel prize was awarded. Indeed, the encomium written by a renowned electron microscopist and published in *Science* on the awarding of the Nobel prize said just that, in spite of the fact that such proof still in great part eluded scientists at that time. Today, he said then, the mechanism of transport has been established. And while it is true that in the more than 25 years since, new evidence, including the first dynamical evidence, has been obtained, the claim that the theory has been proven true still, I believe, rings hollow.

I should take a moment to at least briefly comment on some of the newer evidence before we again turn our attention back in time. Most significantly, exocytosis or some similar event has finally been observed in certain live cells. It has been seen most clearly in mast cells (cells that secrete histamine and serotonin). These observations were made possible by the presence of unusually large secretion granules in a genetically unique strain of mice. In these mast cells it was possible to observe something akin to exocytosis using a relatively recently developed method that enhances ordinary visible-light microscopic images. Although questions about the nature and generality of the mechanism were still left open (for example, they looked more like Heidenhain's granules popping out of cells than exocytosis), these observations provided powerful proof for the existence of a process something like exocytosis in this cell. Other vesicle-mediated events have also been seen in recent years using a variety of new methods of visible-light microscopy, such as confocal imaging, often in conjunction with techniques that label the particular objects being examined, though these observations often seem to raise as many questions as they answer.

There is an interesting irony here. A pioneer in the development of one of these new methods in visible light microscopy was a Professor Robert Day Allen at Dartmouth University. Allen was a student of none other than Lewis Heilbrunn. Heilbrunn's contempt for electron micrographs of dead and altered tissue was apparently shared by his student, who devoted his career to finding a way to provide visual information on living cells beyond the resolution limits of traditional visible-light microscopy.

Other evidence of exocytosis has also been reported. For example, increases in the electrical capacitance of the cell membrane are sometimes seen when secretion is stimulated. Such increases are consistent with the addition of the membrane enclosing secretion granules to the cell membrane, as would be expected during exocytosis. Additional membrane increases the membrane's capacitance. However, the interpretation here is less certain than with the microscopic studies. For example, an observed increase in electrical capacitance could be due to changes in the structure of the extant cell membrane, not additions to it. Most important, thus far this type of evidence has not been quantitative.

Otherwise, in spite of substantial progress, significant proof for equally important elements of the vesicle pathway remain essentially absent 50 years after its proposal. For example, the intracellular transport of products in small vesicles is still a matter of substantial uncertainty, in spite of an elaboration of hypotheses and indirect evidence. To my knowledge their movement has yet to be observed in an intact system, although the movement of molecules thought to be contained within them has occasionally been equated with their movement. But more important, as a general matter, assuming that mechanisms such as those proposed in the vesicle theory do occur, the extent to which they account for transport remains unknown. That is, quantitative evidence has been virtually absent.

This is consequential. Quantitative evidence is crucial in the study of molecular transport. In its absence, even if we knew that a particular transport mechanism existed, we could not say to what extent it accounted for the movement of a substance. That is, vesicle transport might exist all right, but be quantitatively trivial. Or it might exist in one cell, but not another. It might be significant here, but minor there, or be significant under one set of circumstances and trivial in another. Neither the vesicle theory's hegemony nor its claim in one system or another, for one process or another, can be justified without clear and unequivocal quantitative evidence. In its absence, such mechanisms, given their existence, might turn out to play a totally different role than the one envisioned. For example, they might be involved in the recycling of granules and membranes as part of their turnover by the cell, and have little or nothing to do with secretion.

Theories that allow for quantitative predictions, as any theory of transport must, call for quantitative assessments of their validity. Indeed, no theory of mechanism can be affirmed without evaluating its quantitative fit to the process it is thought to explain. But the only way that we can obtain such evidence is by examining the intact process. And this is the core issue. It was not that it was particularly difficult to obtain quantitative data. It was thought unnecessary. The reasoning was that qualitative proof for particular mechanisms like those proposed in the vesicle theory tells us that such a process exists; and these mechanisms have no reason to exist except to carry out the functions we imagine for them. The strong-microreductionist principle relieves investigators of the burden of obtaining quantitative data, but in so doing also prohibits them from obtaining the proof they need.

Making Vesicles

There was also a problem with some of the vesicles and what they represented. When large structures enclosed by lipid membranes, such as whole cells, the endoplasmic reticulum, or the Golgi apparatus, are fragmented, they tend to form small vesicle-like remnants. Such artifactual fragmentation invariably occurs when the cell's contents are homogenized. It can also occur as a result of fixation or other chemical and physical procedures used to prepare tissue for microscopic viewing. That is to say, vesicles either collected from homogenized cells or seen in sections of whole cells may be artifacts produced by the investigator. Obviously, it is important to be able to determine which vesicles are artifacts and which are native structures.

This is of particular concern when we consider the microvesicles that are central to the vesicle theory. How do we know that *they* are not artifacts, fragments of larger cellular objects that we have produced inadvertently? That this can be the case was demonstrated dramatically in one study on the transport of plasma proteins across capillary endothelial cells (the cells that line and form the blood capillaries). It is thought that proteins exit the bloodstream across endothelial cells via small vesicles that shuttle them from the internal (blood-facing) membrane of the capillary cells to their outer surface (facing the interstitial space—the space between the capillary bed and the cells of a particular tissue). Investigators discovered that they could greatly alter the number of small "transport" vesicles (by as much as 75 percent) simply by varying the conditions used for chemical fixation.

The bottom line is simply that we often have no way of knowing whether certain vesicles, particularly small vesicles, that are seen in sections of cells or are found in tissue homogenates are natural structures or artifacts of preparation. Or to the extent that both are present, which is which.

Confusion Between What Is Known and What Is Believed

All of the problems with the vesicle model began with a confusion of reasoning typical of the histological tradition. It was, and remains, an article of faith in the field of microscopic anatomy that function can be inferred from structure. This conviction led many to conclude that the vesicle hypothesis had been proven true by the anatomical evidence and that as such it served as an impressive example of the success of the strong-microreductionist principle. This judgment was based on a mistaken reading of the statement that structure embodies function.

It is self-evidently true that structure embodies function. Where else could one embody it, if not in structure? How could functions exist without the underlying structure of the material world? But a fallacious implication is often drawn from this understanding, namely, that a particular functional embodiment can be apprehended inductively by a scientist observing the structure—for example, when the presence of exocytosis is inferred from the structure of the omega figure. The truth of the matter is that if and when a particular function is discovered this way, it is due to good luck, not good reasoning. Armed only with inductive inferences, the scientist can only generate hypotheses for function from structure.

The idea that a factual determination *can* be made in this manner has often led scientists to believe that proof exists that a structure carries out a particular function when there is in fact no proof. It was in this context that the omega figure was often discussed in frankly dynamical terms, as if investigators had observed the fusion of membranes and the expulsion of granule contents in the microscope when they had seen no such thing. Authors would direct readers to static geometric figures in electron micrographs and talk about events occurring. In point of fact, the investigator had no idea of whether anything had occurred, and there was no doubt that nothing had occurred when the electron micrograph was being taken. Nor could one even conclude that things would look this way if they had. The dynamic essence of the vesicle theory was, and in great part remains, a human artifact—an hypothesis devised by scientists to interpret static structural evidence in dynamic functional terms.

Bottoms Up!

I have argued that the "bottoms up" inductive approach to science that is the strong-microreductionist principle inevitably leads to confusion between hypothesis and evidence. Some would no doubt contend that this determination is mistaken. It can be shown, they might say, that function can be inferred from structure inductively. For example, protein chemists do it every day. They look at the structure of a particular protein of unknown function and infer its function from its structure. If this impression is given, it is incorrect. Protein chemists, or for that matter biological chemists in general, cannot infer function from the structure of a molecule in and of itself—that is, without additional information. And that additional information is nothing less than prior knowledge of its function. Even if we had substantial structural understanding on which to base a particular functional theory, in the final analysis the functional realm itself must be consulted if we hope to explicitly connect an observed structure to some particular function.

That this is so can be seen from the Human Genome Project. As a consequence of this effort there is a very rapidly growing list of proteins of known, or at least partially known, structure, but unknown function. We can obtain certain valuable clues about the function of these proteins by comparing them to the structures of similar proteins of known function. But such clues are just that: hypotheses that

The Making of a Paradigm

then must be tested in the functional realm for each and every particular molecule.

To determine function we must consult the whole system, whether cell or organism. There simply is no way around it. That this is true can be seen from the growing effort to identify the function of protein molecules discovered by genomic exploration in the whole animal or plant. For example, so-called transgenic animals can be made that express, do not express, or improperly express a particular protein of interest. With such animals the scientist then explores how the function of the organism is affected by the manipulation. How else could the function of the molecule be verified except in the whole animal or plant, whatever we might learn in a test tube?

The belief that a rigorous scientific program can be carried out by inferring function solely from the study of isolated structures, whether molecules or anatomical structures within cells, is a false conviction. Knowledge of function does not emerge from structure all by itself. Though structure/function hypotheses are often extremely useful in generating research, they remain hypotheses until and unless they are exposed to the fire of deductive testing against the actual properties of the intact system. It is for this reason that the visible-light microscopic evidence for exocytosis mentioned previously, in which granules were actually observed leaving the cell in real time, was such powerful evidence. It was powerful because it was the first *real* evidence of such a process. Almost everything before it, more than 40 years in the making, was simply hypothesis.

CHAPTER 10

The Experiments

If people will not agree on method, they will not agree on substance.

Alan Wolfe, *The New Republic*, 1997

Readers who are knowledgeable about this subject might complain that I told only part of the story in the last chapter; that I neglected something that is both important and mitigating. They would point out that the scientists who developed the vesicle theory were aware of the dangers of building dynamic functional models solely on the basis of static electron micrographs. They might submit that the scientists' awareness of this peril, and more broadly their awareness of the limitations of the traditional histological approach of inferring function from structure, led them to develop innovative experiments that were intended to overcome such limitations. And it was the results of *these* experiments, they might say, and not the anatomical evidence alone, that gave them confidence that the vesicle theory was correct and that prompted the Nobel committee to award its prize.

But I have not forgotten nor did I intend to ignore these experiments. They were formative experiments of modern cell biology and have been methodological pillars of the discipline. This chapter will be devoted to them. The experiments I am referring to were intended to provide a description of the transport process as it occurred in real space and time; that is, they were intended to provide the dynamic evidence that was missing.

However, as it turned out, though the studies were ingenious and evoked a dynamic mental picture, and were certainly crucial in molding opinion, they did not provide the dynamic evidence they claimed, though it was widely understood that they did. They no more provided dynamic evidence of the events of transport than the omega figure provided dynamic evidence of exocytosis.

Central to the problem was that the experiments were linked to the particular mechanistic embodiments envisioned in the vesicle model. As we shall see, their design and interpretation were dependent upon them, not independent of them. In essence, the investigators had assumed the vesicle theory correct a priori, and sought to verify this assumption by means of their experiments. But if the theory was assumed correct at the outset, then either it was not really a theory, or alternatively, if it was, then the experiments could not provide evidence affirming it. That is, if the mechanisms posited in the vesicle theory could be assumed to occur, then the experiments served no purpose. They would merely reinforce facts already known, already in evidence. On the other hand, if their occurrence was merely presumed, then how could we use experimental results obtained on the basis of that presumption to prove that the presumption was justified in the first place? Surely, the results could not serve as preexisting proof of their own meaning.

But this is exactly what happened. The evidence was used a posteriori to justify the experiment and its interpretation a priori. Because the results obtained appeared to be consistent

The Experiments

with the vesicle theory, the initial presumption regarding the presence of the mechanisms it proposed was seen as having been justified. As a consequence of this linkage, these experiments taught us precious little about the properties of protein transport that was *not* either already known or contingent upon the correctness of the vesicle model. As it happened, the dynamics of the transport process could be said to have been measured *only if we could assume at the outset that the vesicle theory was correct.*

We will see how this peculiar state of affairs came about in this chapter. The manner was subtle, and only implicit. There was no direct statement that the vesicle theory was being assumed true a priori or that the experiments were designed to measure the dynamics of known events. Such a statement could not be justified, and was not made. Nonetheless the presumption was there; implicit in both the design of the experiments and their interpretation. As we shall see, to interpret the results coherently, it was necessary to *disallow* all potential explanations for the observations *other than those proposed by the vesicle theory.*

Front and center amid this presumptive and interpretative perplexity was strong microreductionism. It was the cause of the confusion, as evidence and hypothesis and fact and theory became conflated. Though the *appearance* was that the vesicle theory was being tested against properties of the natural process, in actual fact, the experiments, true to their strong-microreductionist origins, were an attempt to garner evidence on certain *parts* of the cell that were thought to play particular roles according to the vesicle model, while drawing inferences about properties of the *whole* intact process.

The measurements gave the illusion that an intact mechanism was being observed because changes in space and over time were being gauged as they occurred. But in point of fact they provided only a faux dynamism. My intention in this chapter is to show how this occurred, and how the strong-microreductionist viewpoint, hidden in many small experimental details, insinuated itself on global perceptions of what was being scrutinized. The devil is in the details. It is here that implicit beliefs do their dirtiest work.

Two Experiments and Two Incarnations

The experiments in question were originally designed in the 1950s by George Palade, then an electron microscopist at Rockefeller Institute, and a biochemist colleague, Philip Siekevitz, and were an extension of experiments pioneered earlier by others. There were two sets of experiments on the exocrine pancreas, each with two incarnations. The first, performed by Siekevitz and Palade, was carried out in situ (in the living animal). The second was carried out some 10 to 15 years later by Palade and one of his students, James Jamieson, on whole tissue in vitro. The design of both experiments was basically the same. In the in situ study, amino acids, the building blocks of proteins, labeled with a radioisotope, were injected into the bloodstream of animals; in the in vitro study, they were added to the medium bathing pancreatic tissue in a flask. In both cases, the labeled (radioactive) amino acids were taken up by the cells of the gland and incorporated into new proteins as they were being manufactured.

This ability to label the protein with a radioactive substance made it possible in theory to follow its movement. All one had to do was find a way to follow the path of radioactivity, like a beacon, within and out of the cell. Presumably the labeled compound would behave identically to its unlabeled brothers and sisters, and hence would trace their move-

The Experiments

ment. Access of the radioactive amino acid to the synthetic machinery was limited to a brief period of time so that all of the material manufactured before and after remained unlabeled. It was thought that this would make it possible to follow this small aliquot (portion) of material as it moved through the cell toward the exterior, like a wave approaching the shore, being both preceded and followed by unlabeled forms of the same molecules.

At first, we would expect to find the radioactive protein where proteins are manufactured—on and around the ribosomes and the endoplasmic reticulum. Later on, it would appear in zymogen granules, where it would be stored until it was released from the cell. And of course it would eventually end up in fluid secreted by the gland. The experiments demonstrated these particular facts clearly. But they were already known. It was what happens *between* the ribosome and the zymogen granule (and secretion) that was uncertain. How did protein get from the site of its synthesis to the zymogen granule and from there into the duct system? Was the path the one delineated in the vesicle theory?

Attempts to answer this question used two different experimental protocols—prototypes of what were to become standard methods. In the first, pancreatic tissue was homogenized (ground up) at different times after the administration of the radioactive amino acid, and various parts or fractions of cells, such as the zymogen granules, were purified from the homogenate. This was accomplished by centrifugation. Cell parts could be separated from each other by their distinctive sedimentation properties due to differences in their size and density: different parts sedimented at different gravitational forces. The idea was that the timing of the appearance of the radioactive protein in the various separated cell fractions would *mimic* their appearance in these same compartments within the intact cell. In this way the path followed by the digestive enzymes could be delineated.

The second method was electron microscopic autoradiography, essentially the same technique described for muscle in Chapter 7. Autoradiography is complementary to cell fractionation because it examines the parts of the cell *in place*, with the natural relationships intact. There was thought to be great value in using both methods in the same study. The weaknesses of each approach would be compensated for by the strengths of the other. In tandem, it was hoped, they could furnish a reliable understanding of the natural events.

The radioactive isotope used in these autoradiography experiments was tritium, an isotopic form of the hydrogen atom. The tritium isotope was incorporated into amino acids that were then administered to the animal or isolated tissue. As in the muscle study, a photographic emulsion was laid directly over tissue sections. The radioactive protein would periodically emit an isotopic particle that collided with the photographic emulsion and exposed it, leaving a signature spot at a particular location on the film. The source of the spot, or *grain*, could be established by using the electron microscope to find what cell part underlay it. In this way the distribution of grains, and changes in that distribution over time, could be followed as the radioactive protein moved through the cell.

Differential Centrifugation of Tissue Homogenates

We will consider the results of both types of studies; first cell fractionation and then autoradiography. The most fundamental and obvious requirement for a study in which parts of cells are isolated is to be sure that the parts being examined are actually the parts that you think you are examining. That is, the nature of each fraction has to be established to ensure that it contains what is anticipated and that it is not contaminated with other material that could confound the interpretation. Toward that end, Palade and his colleagues used two operations. The first was electron microscopy itself. The various cell fractions were examined with the electron microscope and matched to like structures in sections of whole cells. The second was chemical analysis. The cell fractions were assayed for particular chemical components that were known or thought to be contained in specific cellular structures.

Zymogen granules were relatively easy to identify microscopically in cell fractions because of their striking appearance. Likewise, the presence of ribosomes on the membrane of the endoplasmic reticulum (ER) provided a convenient way of identifying the ER, or at least fragments of it. As a result of homogenization, the ER is broken up into small vesiclelike structures called *microsomes*. The ribosomes are appended to their membranes just as in the intact ER.

Zymogen granules and microsomes were easy to separate from each other because of their widely differing sedimentation properties. Zymogen granules, among the largest and densest cell parts, sedimented at about $1000g$, whereas microsomes are very small and can be separated from the suspending fluid only at centrifugal forces on the order of $100,000g$. As a consequence of this large difference, there is virtually no cross-contamination between the two fractions. The contents of the fractions were confirmed by chemical analysis. Zymogen granules were identified by their high concentration of digestive enzymes, and microsomes by the presence of high levels of RNA (necessary for protein synthesis) or cellular enzymes known to be associated with these structures.

Other cell parts could also be identified visually and chemically in various sediments. There was a nuclear fraction that contained high levels of DNA and could be seen to contain nuclei in microscopic images; and a mitochondrial fraction, with its characteristic mitochondria and their internal membranes, as well as the enzymes of oxidative metabolism (the oxidation of food takes place within the mitochondrion). The chemical identification had a circular quality in that knowledge of the chemical composition of cell parts came largely from cell-fractionation studies in the first place. But in spite of this stipulation, the identification of some cell parts, significantly the zymogen granules and microsomes, was secure.

But there were two difficult problems. The first was that it was not possible to collect all of the relevant fractions in a pure or even relatively pure form. While all fractions were contaminated to one degree or another with other cell parts that sedimented at similar gravitational forces, two structures most important to the vesicle theory, the Golgi apparatus (that is, its vesicular fragments produced by homogenization) and the microvesicles (those small vesicles found near the ER and Golgi) were particularly problematic. Not only could they not be separated cleanly, but they sedimented together, along with a muddle of other small vesiclelike structures of about the same size and density, including fragments of the cell membrane. These objects were indistinguishable from each other visually and at the time were either poorly or completely uncharacterized chemically. As a consequence, one could not obtain sediments that could unambiguously be said to be either microvesicles or Golgi vesicles.

In addition, the empty or filling zymogen granule (the condensing vacuole) could not be separated from the filled or mature zymogen granule, nor could gradations in filling among granules be teased out by their differential sedimentation. They were all deposited together as a single fraction. As a result, the wave of radioactivity could not be followed from its initial appearance in newly forming granules, through stages of granule filling, to the fully filled object, as the vesicle theory predicted.

The inability to obtain these fractions was severely limiting. Critical proof just could not be obtained. What was unknown and what was sought was how the digestive enzymes traveled from the ribosomes to the zymogen granules. But it was just the structures that were alleged to be involved in this movement that could not be recovered as clean, separable fractions. Did the proteins move in the small vesicles? Did they pass into and through the Golgi apparatus? Could the gradual filling of initially empty zymogen granules be established? These central questions could not be answered.

But there was another shortcoming that was even more hobbling. Once the cell had been broken open by homogenization, the labeled protein had the opportunity to be redistributed among its various parts. In this case, how could one be sure that what was measured in a particular fraction had been the substance's natural home in the intact cell? It may have been added, in whole or in part, from some other source. Most devastatingly, how could one ever establish the occurrence, much less the extent and character, of such a shift?

The act of homogenization had introduced a crucial uncertainty. The question of the relationship between the location of the material in the various cell fractions and its location in the intact cell could not be settled with any assurance. And yet this information was needed if the movement of the labeled protein from place to place was to be determined. We have come across this problem before. It is the inevitable result of the loss of the entailing whole. In this case, its loss was a consequence of destroying the natural relationships between the parts by homogenization.

Even though the experiment was designed to establish the movement of proteins in the intact cell, before any measurement could be made it was necessary to break open the cell and thereby destroy the intact structure that the investigators sought to examine. Although this is an important, indeed a limiting, problem for all cell-fractionation studies, it does not mean that tissue homogenization is of no use in experimental biology or that any information derived from such measurements should be ignored. Breaking open the black box that is the cell has been crucial to understanding it chemically. Without this "destructive" act, our understanding of the chemistry of life might well have remained essentially as primitive as on the day cell theory was proposed. But if our expectation is to fully understand intact cellular mechanisms in this fashion, mechanisms that have spatial entailments, we are doomed either to failure or self-delusion. By the act of homogenization we lose crucial information that simply cannot be recaptured by the force of the human intellect.

How the ambiguities produced by homogenization were handled in the case of the vesicle theory is instructive and can serve as an example for similar interpretations made in comparable fields of study. More generally, it is illustrative of a common approach to the study of the cell. The sad fact was that major assumptions had to be made in order to interpret the results. Specifically, the investigators had to presume the particulars of redistribution subsequent to homogenization!

What were these presumptions? First, and most innocuously, it was presumed that some redistribution of labeled protein from one unspecified fraction to another must have occurred. That is, it was held that something had moved. This seemed fair enough, even obvious, because of the massive disturbance produced by homogenization, but of course was of no help in interpreting the results. It was necessary to presume something much more specific; in particular, explicitly and quantitatively, what had been redistributed, from where to where? This was done and it was taken as known that certain cell fractions were quantitatively contaminated from other sources, that is, they contained *only* contaminants.

Which fractions were thought to be contaminated and which were not depended on their relationship to the vesicle theory. If the presence of the labeled protein was predicted by the theory, then it could be assumed to be in its natural home, essentially uncontaminated by material from elsewhere. If it was not, then it could be assumed to be an artifactual inclusion. And finally, and most interesting, if material was not present in a fraction where the vesicle theory predicted it should be, or was present only in small amounts, it was assumed to have been lost as a result of homogenization. Such assumptions led to the systematic discounting of all evidence that did not fit the vesicle theory, while affirming evidence that appeared to validate it.

The Cytoplasm

The most important of these assumptions was that any labeled protein found in the nonsedimentable fraction—that is, the supernatant or soluble fraction of the homogenate which contains the cytoplasmic contents of cells—had been redistributed from other sources. It seemed reasonable enough to assume that some of this material came from other cellular compartments that might contain digestive enzymes due to their breakage during homogenization—for example, from damaged zymogen granules. But the presumption was not that *some* of the content of the supernatant or cytoplasmic fraction was due to an artifactual redistribution, but *all* of it.

This conclusion was of great weight. If it was justified, it meant that the secreted proteins were not normal constituents of the unperturbed cytoplasm. And if this was true, then so was the vesicle transport theory! That is, if the digestive enzymes were excluded from the cytoplasm, how else could they move through the cell except from one membrane-bounded intracellular compartment to another, eventually exiting the cell from one of them, much as the vesicle theory proposed?

If, on the other hand, the proteins *were* constituents of the cytoplasm, the vesicle theory could not be wholly—that is, quantitatively—correct, and indeed it might be totally false. At the least, it would mean that another, nonvesicular, pathway existed. Distinguishing between these two possibilities would have been a significant test of the vesicle theory. Was there or was there not a nonvesicular or cytoplasmic pool of secreted product? If it could be shown that none existed, then exclusive attention could properly be focused on transport by vesicles. On the other hand, if this could not be shown, the vesicle theory would have to be rethought.

But this challenge was not taken and perhaps not even appreciated. It seemed to be satisfactory to *assume* that labeled protein in the supernatant fraction was contamination—an artifact of redistribution. Radioactive protein was routinely found in the cytoplasmic fraction, and its presence as routinely discounted, thus immunizing the vesicle theory from falsification.

Perhaps the nature of the bias at work can be made clear if we imagine the interpretation that might have been made if radioactive protein had *not* been recovered in the cytoplasmic fraction—that is, if it were free of labeled protein. If this had

The Experiments

been the case, it would quite naturally have been seen as very strong evidence for the vesicle theory. In all likelihood, it would have been argued forcefully that the absence of labeled protein proved not merely the occurrence but the *dominion* of vesicle mechanisms. But couldn't the absence of labeled protein in the cytoplasmic fraction also be an artifact, reflecting its *loss* into particulate fractions (due to protein adsorption or diffusion)? Indeed, the nonionic medium commonly used for homogenization favors protein adsorption to membrane surfaces.

In any event, without excluding the cytoplasm presumptively, the cell-fractionation experiments told us little more than we already knew. We knew that new protein moved from the ER to the zymogen granules. The question was not whether this occurred, but how. Did passage occur by means of small vesicles, through the Golgi apparatus, into condensing vacuoles? Or alternatively did new protein diffuse through the cytoplasm of the cell prior to storage in the granules? Or did both occur? After all was said and done, the experiment did not illuminate the events involved, much less illustrate the presence of the mechanisms proposed by the vesicle theory.

Microsomes

But there was another observation that gave investigators confidence that things occurred much as the vesicle theory proposed. It involved the endoplasmic reticulum and its homogenized fragments, the microsomes. In the homogenization studies we have just discussed, the amount of labeled protein associated with the microsomes was measured at various times after the addition of radioactive amino acids. An attempt was made to distinguish between labeled material on the ribosome itself and within the internal spaces of the microsomal vesicle. The latter was thought to be akin to the internal spaces of intact sacs of endoplasmic reticulum within the cell.

As described earlier, it is widely thought that the digestive enzymes enter this space directly from the attached ribosomes during their synthesis. As the peptide chain is being synthesized, elongating one amino acid at a time, the new, more or less linear chain snakes its way through a pore in the membrane directly under the ribosome's point of attachment. This is called *cotranslational translocation* to specify that the protein is moved during its synthesis, as the nascent chain is being translated from its RNA code, and significantly not afterward, when it is folded into its mature, more or less globular or spherical form.

To obtain the microsomal vesicle's contents, its fatty membrane had to be solubilized with a detergent. This released the ribosomes as well as the vesicle's contents into the suspending medium. The two could then be separated by sedimenting the ribosomes in an ultracentrifuge at very high speeds. If it could be shown that there was a shift in the distribution of the newly manufactured protein from the site of its manufacture, the ribosome, to the contents of the microsomal vesicle, the detergent-soluble fraction, this would prove that transport had occurred from one to the other as the vesicle theory proposed.

Such measurements were also made in another way. In this case, microsomes were separated from the homogenate and suspended in appropriate media *prior* to the addition of labeled amino acids, rather than afterward. That is, labeled amino acids were added to isolated microsomes, not whole cells. This allowed the process to be studied at the level of the microsome itself. Amino acids could be incorporated into

The Experiments

proteins in such isolated preparations, and the fate of newly synthesized protein followed in the same fashion as in the experiments on whole tissue.

In both experiments, more labeled protein appeared in the detergent-soluble fraction at later times, seemingly confirming the hypothesis that new protein was being transported into the internal spaces of the microsome (in the study with isolated microsomes), or into the cisternal spaces of the endoplasmic reticulum (in the experiment with intact cells). If this was true, the weaknesses of the cell-fractionation studies would seem less significant, if not wholly irrelevant.

The reasoning went something like this. If some new protein manufactured on the attached ribosomes entered these internal spaces, then all protein manufactured on them did. That is, qualitative proof for the existence of such a transport process could be taken as quantitative proof for it. Further, it was assumed that having entered these spaces, the protein molecules would be held there irreversibly, unable to leave by diffusion back across the membrane. In the jargon of the field, they would be *sequestered*.

If one could have confidence in both of these assumptions, quantitative transport and sequestration, then the proteins in question would necessarily be wholly excluded from the cytoplasm. If all entered and none left, there could be no cytoplasmic pool. And having thus excluded the possibility of a cytoplasmic route, the vesicle model provided the only possible means for the movement of proteins out of the ER and through the cell. They had to leave the ER in one way or another if they were going to be secreted, and it seemed that the only way that this could happen was by means of vesicles that bud from reticular membranes. And likewise the remainder of their travel through and out of the cell would have to be by means of one vesicle or another.

But the evidence showed none of this. At most it showed that *some* new protein ended up in the internal space of the microsomes. The determination that all protein entered this space and that none left was merely a supposition. Why, it was asked, would some molecules be transported into this space if they all weren't? And why would protein be deposited in the cisternae of the ER merely to diffuse out again? Indeed, why would molecules enter at all, unless something very much like the events proposed in the vesicle theory took place?

As it happened, the data did not even allow for the unambiguous conclusion that new protein had entered the internal spaces of the ER, much less all of it and irreversibly. To draw this conclusion, an additional assumption had to be made. It had to be taken as known that the detergent-soluble material contained only the contents of the vesicles, nothing else. But this was clearly incorrect. It contained all other material from the microsome that was released into the soluble phase upon the addition of the detergent. This included detergent-soluble constituents of the vesicle membrane; detergent-soluble material previously adsorbed to the membrane or to the ribosomes; and substances in the aqueous (watery) phase of the microsomal sediment external to the microsomes themselves that were released simply as a result of the sediment being resuspended in new medium. Not only could it not be concluded that all of the labeled protein in the detergent-soluble fraction came from the internal spaces of the microsomal vesicles, it was possible that none of it did.

In the experiments done on whole cells, there was an additional problem. In this case, not only might the detergent-soluble material have come from some other microsomal location, it might have been *added* or trapped within the internal spaces of the microsomes as they were being formed during homogenization. How could one be sure that what was found inside them was not at least in part an artifact of *inclusion,* something added during homogenization? In fact, how could one tell that any material was native to this compartment of the intact cell, if all one examined was the contents of microsomes after fracturing the ER during tissue homogenization?

This possibility was not tested during the years of the vesicle theory's development. It was simply assumed that the natural contents of the cisternae, and those contents exclusively, were being measured. When this conclusion was examined many years later, the results were disturbing. An exogenous material, a large polar polysaccharide, inulin (not the hormone insulin), was added to the suspending medium *prior* to tissue homogenization. The idea was to see if inulin was recovered in the detergent-soluble fraction of the microsomes, and if so in what amounts.

If inulin was not present at all, or only in small quantities, one could conclude that not much trapping had occurred as the microsomes were being formed, and therefore that the presence of the labeled protein was in all likelihood a reflection of the native contents of the ER. On the other hand, if inulin was found in amounts comparable to those seen for the labeled protein, then significant trapping of labeled protein was a distinct possibility. When this experiment was performed, the inulin was not only found in the detergent-soluble fraction of the microsome, but at roughly the same percentage (of the total in the homogenate) as the labeled protein. Thus, apparently there had been substantial trapping of material, and the presence of labeled protein in the detergent-soluble fraction might be—and might even *solely* be—an artifactual addition. And if this were not enough, about 75 percent of the labeled protein could be removed from the microsomes prior to the addition of the detergent simply by washing them in water. This meant that most of the protein content of the detergent-soluble fraction came from the external medium and not the internal spaces of the vesicles, was adsorbed to the surface of the microsome, or was otherwise able to diffuse out of the vesicles.

All in all, the evidence for transfer into the ER and sequestration in its internal spaces was loaded with unexplored suppositions. Not only that, the evidence actually supported the existence of a cytoplasmic pool of digestive enzymes. Most simply, as said previously, a significant amount of digestive enzyme was found in the cytoplasmic fraction of homogenates. Although the amount varied substantially, from about 15 to 50 percent, depending on the particular protein molecule and animal species, it was invariably found.

How could one exclude the possibility that at least some of the new protein was released directly from the ribosome into the cytoplasm (not the ER)? In fact, the studies on isolated microsomes that were thought to consolidate the vesicle theory's perspective seemed to demonstrate just that. A substantial amount of labeled protein, approximately one-third to one-half as much as that found in the detergent-soluble fraction, was released into the *medium* bathing the microsomal vesicles as they were manufacturing new product. Since only microsomes were present in the medium, the labeled protein necessarily came from the ribosomes, either directly or indirectly by passing back out of the vesicle into the medium. Either way, the evidence did not affirm the absolutism of the sequestration hypothesis. To the contrary, it provided substantive evidence of a cytoplasmic pool of digestive enzymes.

The Experiments

In the many years since the original studies were carried out, studies purporting to show the cotranslational translocation of nascent protein have been legion. Almost universally, the crucial compartmental distinctions we have been discussing were ignored, assessed indirectly, or assumed to have already been established. More important, given the occurrence of such a process, it was assumed to be quantitative; to apply to all molecules of a given protein.

Autoradiography

Thus, if labeled protein was found in a favored compartment, its presence was assumed to be natural. If it was found elsewhere, it was assumed to be an artifact. All that seemed to be required to discount an observation that did not fit the vesicle theory was to be able to imagine an artifactual explanation for it. There seemed to be little concern that such assumptions could be wrong. Nonetheless, there must have been some awareness of the interpretative problems with the cell-fractionation experiments, because a complementary approach—electron microscopic autoradiography—that offered the opportunity, or so it seemed, to overcome at least some of them was also employed.

As explained, in autoradiography cells are viewed in microscopic sections of tissue with the natural relationships between their internal parts more or less intact. In this case, the movement of the labeled protein is tracked by examining tissue sections with an electron microscope and locating the emitted autoradiographic grains on a film emulsion laid over the tissue section. Samples were examined at different times after exposure to the radioactive amino acid, just as in the cell fractionation experiments. However, in this case the location of the labeled protein was assessed by tracking the autoradiographic grains.

Autoradiography seemed to avoid many of the pitfalls and ambiguities of the cell-fractionation measurements. Quite simply, the location of the grain identified the location of the protein. It was either here or there. And if the grains were found over the expected compartments in the expected sequence, the vesicle model would be validated.

But as it turned out, things weren't this simple. Like cell fractionation, autoradiography proved to be a very complex method with many uncertainties of measurement and interpretation of its own. Most important, it turned out that one could not resolve the location of the protein simply by looking for grains. In autoradiography two methods are blended—microscopy, in this particular case electron microscopy, and the use of radioactive isotopes to trace the movement of substances. That this approach can be useful was shown by experiments such as those describing calcium's role in muscle contraction, but bringing together electron microscopy and the use of isotopes to measure molecular movement introduces a variety of difficult technical and interpretive problems. Nor did the merger alleviate any of the problems normally inherent to electron microscopy. In fact, in certain ways it made them more acute.

Methods of Preparation

To understand the meaning of results obtained using autoradiography, we have to first understand the technical limitations of the method. Some of these are limitations of electron microscopy or microscopy in general, while others are specifically related to the autoradiographic application. So before discussing the experiments themselves, we will consider some of these technical issues—in particular, the fidelity of the sample to the native state; the problem of adequately sampling

The Experiments

very small things; the effects of fixation on the sample; the limits of resolution; and finally, how we develop the photographic end product. Each can have a great impact on what we see and what we think it means.

First, fidelity. Central to the utility of autoradiography is the preparation of the sample for viewing. As already discussed, in ordinary electron microscopy, we have to be concerned about whether the image we conjure is faithful to the natural object. In autoradiography, the stipulations for fidelity are, if anything, more demanding. As a general matter, the development of methods of sample preparation for various forms of microscopy has been a heuristic enterprise. By this I mean that the procedure is usually considered successful if it addresses central practical concerns.

For example, as we have discussed, the sample often has to be cut into thin sections to make it transparent enough for the cell's internal structures to be seen. To accomplish this task, other procedures are usually needed. For instance, in order to cut or section soft biological tissues, they must first be embedded in a relatively rigid matrix. To do this, matrix materials must fully infiltrate the tissue. And to achieve this, the major component of the cell—its water—must first be replaced by an organic solvent. Thus, to cut thin sections, a variety of treatments are required—the sample must be dehydrated, and its water must be replaced with an unnatural solvent and then with a rigid embedding material. Only then can we see the internal structure.

Another example is imaging contrast. As noted earlier, all cell parts are of roughly similar density. As a consequence, it is often hard to distinguish one from another in a microscopic image. We say that contrast between them is inadequate. So, one of the microscopist's tasks is to improve contrast so that we can better distinguish various objects. This is usually done by trying to enhance the appearance of a feature by adding chemical stains to increase its density or give it a particular color so that it stands out from others.

Thus, good practice in the preparation of samples for microscopy has largely been assessed by whether the method is successful in overcoming such practical problems of viewing, and not as a consequence of its *faithful rendering* of the natural object. This is not merely an historical fact. There is no choice. With the exception of samples that can be examined alive and thus in their native state, the fidelity of samples that require preparation prior to viewing an unknown natural object is necessarily unknown itself.

Even if we limit our manipulations to freezing the sample to obtain the necessary rigidity for sectioning, we cannot guarantee the fidelity of structure and location. Artifacts are easy to induce by freezing, and investigators often go to great lengths to try to minimize them. Even if we could produce a faithful rendering this way, it would not be enough. We would still have to see the objects clearly, and nature does not make this easy, because untreated material shows very poor contrast in electron micrographs.

And yet the central mission of the biological sciences is to comprehend the natural state of living things. As a result, distinctions of some sort have to be made between different methods we might use to prepare samples for microscopy. For example, say two methods allow us to section material successfully, and both demonstrate enhanced contrast, but they provide significantly different representations of the object we are interested in, leading to differing conclusions about it. We cannot view both as equally good or equally bad depictions of reality, because they provide different images and lead to different conclusions. How then can we determine in a truly unbiased and objective fashion which provides

The Experiments

the more natural result? There is usually no unambiguous way to do this. As a consequence, some sort of value judgment needs to be imposed if we are to learn anything. The unadorned fact is that, as said, most things that microscopists look at can only be viewed contingently—that is, as models of nature. Still, as a utilitarian matter, we must assume that one model, not another, more closely reflects the natural state.

This choice is usually made by consensus or authority, most commonly by a combination of the two. Experts, or at least most experts, come to agree that a particular method gives "good results" while some other approach is less satisfactory. This agreement is often focused on sensible considerations, but whether they are sensible or not, a particular method comes to be favored and is presumed to provide images that are more reflective of the natural state, with fewer artifacts or distortions than those produced by less favored approaches. The conclusion that a particular method of sample preparation is good practice is part of the *art* of science.

Perhaps we can appreciate this problem with an artistic example. A considerable, but usually unstated, criterion for the fidelity of microscopic appearance is aesthetics. As inscribed on Keats's famous Grecian urn, "Beauty is truth, truth beauty." However difficult a proposition to prove, and one that Keats himself did not hold, the conjunction between beauty and truth is often assumed in biology. Methods that produce forms that are visually pleasing are likely to be judged more reflective of the native state than those that are not. Of course, perceptions of beauty vary widely among individuals and cultures, but in biological microscopy, the criteria are usually those of so-called classical beauty—that is, the beauty of the geometric form. Structures and relationships that are less regular, less symmetrical, or at least less geometrically obvious are more likely to be seen as being artifactual than those that have a regular and obvious symmetry. There is good reason for this, because an irregular appearance suggests disorder, and disorder may reflect entropic tendencies in the method of preparation. Geometrical regularity, on the contrary, suggests order, and the likelihood of order retained whatever the method used. But such appearances can be deceiving. Impressive order and organization may underlie an irregular, disorderly appearing object. Conversely, order can be produced artifactually—for example, when normally soluble substances crystallize.

There can be no doubt that we have discovered a great deal about the microscopic nature of the biological world over the past 150 years, not to say the past 300 years since the invention of the microscope. And this is not in spite of such problems, but because of them. That is, progress has often been due to the application of these uncertain preparative procedures. It is through their development that we have come to understand cellular structure. As a consequence, in spite of the essential and unavoidable uncertainty, we have good reason to be confident that the image of the cell that has emerged over the years bears a fair resemblance to the actual thing, and this is a great accomplishment.

But a fair resemblance does not mean a certain and assured identity. And significantly, as we shall see shortly, the more demanding our requirements—in particular, the smaller the objects we examine and the greater the accuracy of location we seek—the more circumspect we must be. When our demand is to know the locations of single molecules in cells, as in autoradiography and other similar methods, we must apply great care in interpreting what we see. As we look closer and closer, at higher and higher magnification, at higher resolution, the problem of fidelity becomes increasingly severe, because only very small alterations are needed to produce artifacts.

The Experiments

In electron microscopic autoradiography, just as with cell fractionation, we have to assess whether the molecule's location as we see it in the autoradiogram is the result of a redistribution from elsewhere. Also as in cell-fractionation studies, we have no means of determining the original—that is, natural—location of a substance of interest, such as a digestive enzyme, *prior* to employing the preparative procedures. We cannot know if and how things have changed if we do not know how they were before they changed. And thus, like tissue homogenization, autoradiography is vulnerable to redistributive uncertainties.

Because we can never know the initial (native) state of the tissue prior to our manipulations, we can never be totally confident that we have ensured its retention, however we seek to do so. And as we will discuss shortly, when some change is inevitable this becomes a matter of great significance, if all that must occur to completely alter our interpretation of what has taken place is a small deviation from the natural state. As we shall see, this was a substantial problem in attempts to explore the vesicle model using high-resolution electron microscopic autoradiography.

Sampling

Another problem of microscopic observation is sampling. In tissue-fractionation studies, we look at a whole organ or tissue, comprising billions of cells, and the information we obtain represents a grand average of the properties of them all. If our measurements are reliable and we take into account variability from animal to animal, we can have some confidence that the values we obtain reflect those for the population of all cells of this particular type.

In microscopy our vision is far more limited. We can examine only a few cells—hundreds or a few thousand, not millions or billions. This has been an historical problem with visible-light microscopy, and the scientific literature is replete with examples of differing perceptions among investigators about the appearance of a structure or the organization of a particular cell, due to the inability to obtain a sufficient statistical sense of what is common, as opposed to what is not. Pictures may not lie, but by viewing different samples, different observers may come to very different conclusions about the nature of their material.

As much of a problem as this has been for visible-light microscopy, it is far more severe when we turn to the ultra-high-resolution electron microscope. As we demand higher and higher resolution, and seek smaller and smaller details, the number of cells that we can examine as a practical matter becomes increasingly small. This is the price that nature exacts for letting us look more closely; and for structures as variable and complex as biological cells, it becomes even more problematic to form satisfactory generalizations. Rarely are more than a few thousand cells examined, and not uncommonly far fewer. It is from this tiny sample that we attempt to generalize about the nature of the object. Although there are sophisticated statistical procedures that can help us deal with this problem, and automation and computers theoretically permit the examination of large samples of small objects, these procedures are infrequently used. Without such methods and without statistical validation, what is reported can only be a sample selected by the observer, using whatever criteria the observer thinks appropriate. As such, and however honest the choices, published images, including autoradiograms, are necessarily vulnerable to observer bias, or simply observer misperception.

The Experiments

Fixation

The first step in preparing samples for traditional viewing with either the electron microscope or the visible-light microscope is *fixation*. The purpose of fixation is to try to overcome the dangers of preparative artifacts like those we have been discussing. Although it is a chemical procedure, like cold temperature it is meant to freeze the sample in its native state, ensuring that whatever additional steps are undertaken will not change it. Commonly, covalent bonds are formed between the fixative substance and various molecules in the cell, such as proteins, forming a chemical bridge that interconnects much of the cell's contents in an intricate meshwork that prevents its various and sundry molecules and structures from diffusing away from their natural or initial locations.

For a fixative to be effective it must act rapidly—that is, before the movement of cellular materials can occur. To do this it must enter the sample more rapidly than disordering can take place. But as the size of the structure of interest becomes smaller, at the limit for individual molecules, even rapidly diffusing fixatives cannot enter quickly enough to keep things precisely in place. Consider a molecule sitting in the center of a cell that sits in the center of a small piece of tissue. In order to reach this molecule, the fixative must not only diffuse through the cell to its center—say, 5 μm—but because fixation must occur before the piece of tissue is sectioned (immediately after its removal from the organism or even before), it has to move through the extracellular and cellular spaces of the larger tissue mass (say, 1 mm or so in diameter). Even if fixation is achieved in situ, through the blood circulation, thereby greatly reducing the distance for diffusion, that distance is still on the order of 25 to 50 μm.

Compare this to the distance that our endogenous molecule in the cell's center must diffuse in order to confuse us as to its initial location. Unless other forces (say, binding to a membrane) constrain or prevent its diffusion, all it need do is move a distance equivalent to the thickness of a biological membrane, about 0.01 μm, out of one and into another compartment, to lead us astray. Even if the cellular substance diffuses far more slowly than the fixative, the difference in the time available for their movement, the one to maintain order and the other to dissipate it, would still be orders of magnitude. Add this to the fact that the amount moved decreases greatly as the distance to be traveled increases, and you can see that expecting a fixative to ensure that molecules do not move away from their native home in the cell is not something about which we can have a great deal of confidence.

But even if we could accomplish this feat, we still would have no assurance that things were left unaltered. Indeed, the idea that fixation reactions that seek to link all of the cell's contents nonetheless leave it unperturbed is counterintuitive. How can chemical bonds that are not natural in themselves and that produce such a massive alteration in the object leave it unaltered? Commonly the intent of such methods is not to prevent all possible change, but to limit it to distortions *beyond* the resolution limit of the method, to changes that are too small to be seen in the microscope. That this is the case when we seek accuracy at the dimensions of molecules, as in electron microscopic autoradiography, is far from ensured.

Resolution

While the issues we have discussed to this point are problems for all sorts of microscopy, the following two are *specific* to autoradiography and the particular experiments we are discussing. The first concerns resolution. As I have explained, grains on autoradiographic photographic emulsions result from exposure of the film to radioactive particles emitted from the tissue sample at a particular distant point in space. As it turns out, the relationship between the location of the source and the location of the grain on the emulsion is not linear. That is to say, one is not perfectly superimposed on the other. Most grains are found at points laterally remote from the emitting source. As a result, they may be found over objects that are not the same as those that actually emit the radiation—that is, that are *not* its source. So how can we determine whether a particular grain is from the underlying object or from elsewhere?

This is dealt with by making statistical estimates of location by examining many grains. Though we cannot specify the exact coordinates of any single grain, it is possible to determine whether the emissions are *likely* to have come from this or that location in the cell. To make this determination, we have to calculate the extent to which an emission from a particular type of isotope—say, of calcium or hydrogen, as in the current study—is likely to stray from the straight and narrow path of superimposition. That is, we need to estimate the degree to which a particular type of emitted particle is likely to move sideways as it moves toward the film. This varies from isotope to isotope as a function of the energy of the emitted particle. And this, in turn, is a measure of the accuracy or resolution of the method. The greater the potential lateral movement of the isotope, the less the resolution of the method.

As the emitted particle moves away from its source, it follows a random path that can diverge at any angle from a straight line perpendicular to the plane of the sample (superimposition). It may even move horizontal to that plane. As the particles move away from the source, they fan out, and such lateral divergence increases with distance in a conelike manner. Hence, the greater the distance the film emulsion is from the source of radiation, the greater the potential deviation. The base of this cone can be imagined as a circular area on the film emulsion. This is called the *resolution element*. It defines the region in which radiation emitted from a particular point source is likely to diverge before hitting its target (the emulsion). If we know the distance traveled (that is, the thickness of the sample and how closely it is apposed to the emulsion) and the energy of the emitted radiation (higher energy particles can move farther away from the source), we can calculate the dimensions of the resolution element, the diameter of the cone's base.

If the resolution element is smaller than the object in which the molecule is thought to reside, then we can be reasonably certain that a grain over that object comes from it. On the other hand, if the resolution element is larger than the object, then the emission may come from other nearby structures or simply from the spaces outside it, and not from the object itself.

What was the size of the resolution element relative to the size of the anatomical objects thought to be the source of the radiation for the autoradiography experiments we are discussing? The answer is that it depends on which object we are talking about. For example, for the average 1-μm-diameter zymogen granule, the circle of resolution for the tritium isotope falls well within its borders. Therefore, grains found directly over a zymogen granule are most likely, though not necessarily, emitted from it. (If found on the edges of the object, they could be from a small region just outside it.) For smaller structures, such as the 40-nm-diameter small transport or microvesicles or the narrow spaces of the cisterns of

The Experiments

the ER or Golgi apparatus (about the same diameter as the small vesicles), the source of the radiation cannot be unambiguously attributed to these objects. That is to say, it may come from other nearby structures in the field of view or from the space outside them (that is, from the cytoplasm).

For example, grains found over the ER cannot be attributed to its cisternal spaces, as opposed to other nearby structures, such as the attached ribosomes, the external surface of the enclosing membrane, or the cytoplasm. The radiation might come from any and all of these locations. All that can be said with any certainty is that the grain comes from this particular *region* of the cell—the region of the ER. Though the term *region* was often used to express this ambiguity, the clear implication was that the grains reflected the contents of the internal spaces of the particular structures of interest, such as the ER, the Golgi, or the small vesicles—a determination that in point of fact could not be made.

It was argued that if the vesicle theory was correct, we would expect to find the isotope in the region of the ER and the Golgi, and this was clearly the case. Even if the results did not provide proof positive that the labeled protein was actually within these structures, they were certainly consistent with such a conclusion and hence supportive of the vesicle theory. But where would we expect to find the labeled protein if the vesicle theory was *incorrect*? As it happens, in essentially the same places. For example, if ribosomes manufacture protein, and are attached to the ER, then the protein would be found in the region of the ER regardless of what happened to it subsequently. And likewise for the region of the Golgi. We would expect to see the labeled protein in this location even if it only passed nearby as it was wending its way from ribosomes at the cell's basal end to the zymogen granules.

Thus, whether the labeled protein was in the region of the ER, the Golgi, or the small vesicles meant little in and of itself, because its presence in these vicinities was to be expected *whatever* the mechanism of transport. It turned out that even though the results were reported as corroborating the vesicle theory, in point of fact they were as much refutation as corroboration. To say that the labeled material was in the particular structures specified by the theory was just another hypothesis masquerading as fact. The material might be in them all right, but on the other hand it might not. And most consequentially, it might be in the cytoplasm surrounding them.

Exposure

The second autoradiography-specific uncertainty concerns the need to choose a particular exposure time for the emulsion, just as with any ordinary photo film. In this case, the number of grains on the emulsion will vary with the time of exposure. The longer the exposure, the greater the opportunity for emissions to land on the film. Ideally, we would like to choose an exposure time that allows for the measurement of each and every source of radiation once, and only once. Though this would be the "correct" exposure, it is not possible.

For one thing, the method permits only a statistical estimate of the emitting sources in the tissue, not their actual enumeration. This is because the emission of radiation from radioactive substances occurs randomly. As a result, even if we were lucky enough to get the exposure just right and measure the true number of emitting particles, in all likelihood we would have counted some more than once and others not at all. If we increased the time of exposure to ensure that each emitting particle was measured, we would

The Experiments

inevitably count some sources multiple times, overestimating their presence.

More important, to choose the correct exposure time, plus or minus this statistical error, we would have to *know* the number of emitting particles in the sample. Only then could we estimate how long an exposure was required to obtain a proper reading. Otherwise we would be ignorant of whether we had undercounted or overcounted the source proteins. In the case of the experiments we are discussing, not only is that number unknown to us, but it is what we are trying to determine by counting grains. As a result, in autoradiography as commonly performed, the time of exposure is chosen somewhat arbitrarily. And as we shall see in a moment, that time is based on what one anticipates seeing. Thus, unless graced by chance, whatever time we choose either overestimates or underestimates the signal, either not measuring all the sources of radiation or measuring some multiple times, or perhaps both. But more significantly, we have no way of knowing either the direction or the extent of the error. That is, we have no way of knowing whether we have overexposed or underexposed the tissue sample, or by how much.

As much of a hindrance as this is, it is not the most limiting aspect of autoradiographic measurements. The uncertainty about exposure time does not merely apply to the whole sample, but to aspects or compartments within it as well. For example, a particular development time may underexpose one compartment in a cell—say, the mitochondrion—while overexposing another. Whether a compartment is overexposed or underexposed depends on the concentration of the emitting particles in that particular niche. If the radiation source is present at a low concentration, it might not be detected at all (underexposure). If we try to rectify such a situation by additional exposure, we run the risk of overestimating the signal from adjacent pools that contain the material at high relative concentrations, in this case producing multiple emissions from the radiation sources within them (overexposure). Moreover, as the total number of grains on the emulsion is increased, we increase the possibility that grains will be found over neighboring compartments as artifactual inclusions, and even counting grains may become difficult as they start to overlap on the film.

As a consequence of all of these limitations and uncertainties, the time of exposure that is chosen is usually based on various suppositions about the sample—as mentioned, what we expect to see—as well as certain practical considerations. As for the latter, common practice has been to choose an exposure short enough to avoid a dense grain pattern that is hard to count, while still providing a significant signal over compartments that contain the radioisotope at the highest concentrations. Though this is helpful, it reduces the sensitivity of the method and biases the results *against* compartments that contain the substance at relatively low concentrations.

Still and all, when few if any grains are seen over a compartment after such an exposure, it is assumed *not* to contain the substance of interest. But how can we know whether this conclusion is correct, or whether it reflects the fact that the exposure time was too short to make an adequate measurement for the concentration of the material present? Not surprisingly then, autoradiography tells us best about locations that contain high concentrations of a radioactive source, but may mislead us in a negative fashion about its residence when it is present at low relative concentrations. The appropriateness of accepting such exclusionary measurements is bolstered by the traditional view that proteins have a unique or singular home in a cell, and that secreted proteins have the various homes specified in the vesicle pathway. A mitochondrial protein, for example, should be found in the mitochondrion, and in abundance, but nowhere else; a nuclear protein should be in the nucleus, but not elsewhere; and a secreted protein may be in a secretion granule or any of the various compartments in the vesicle pathway, but not elsewhere.

When high concentrations are found in the specified compartment or compartments, this seems to validate the a priori assumption about compartmentalization *a circulare*. Today we know from other lines of investigation that proteins are commonly found in more than one cellular compartment. For example, mitochondrial proteins are found not only in mitochondria, but in the cytoplasm, albeit at low concentrations relative to the mitochondrion. Low concentrations or not, their presence in the cytoplasm is not an artifact, but crucial, because to reach their prospective home they must diffuse from the ribosome where they are made through the cytoplasm. The same can be said of proteins in chloroplasts, the nucleus, peroxisomes, and a variety of other intracellular structures.

The bottom line is that while numerous grains over a particular structure within the resolution element is prima facie evidence that that structure is a natural home for the substance being studied, the converse, few grains over a structure, does not prove that that compartment does not contain the substance, though this has often been the interpretation.

When a few grains are found, their presence is usually assumed to be due to background radiation or outliers from nearby high-emitting areas. That is, their presence is assumed to be noise of one sort or another, not signal. But a relative sparcity of grains over a particular structure does not even demonstrate that the amount there is small. That amount is a function of the size of the compartment as well

Lessons from the Living Cell

as the concentration of material in it. A large compartment may contain substantial amounts of a substance at a low concentration, while its presence at a high concentration in a much smaller compartment might reflect less material overall. For example, mitochondria commonly occupy a small percentage of cell volume, say for purposes of discussion 2 percent, whereas the cytoplasm is usually the largest compartment in the cell, on average about 50 percent of the total. Hence, everything else being equal, the amount of material in the two compartments would be the same when the concentration (grain density) in the cytoplasm was 25 times less than in the mitochondria.

And so, what on the surface seems to be a relatively straightforward method, counting grains at various locations within cells, is not straightforward at all. There are all sorts of uncertainties, all sorts of ambiguities, some that can be avoided, but many that cannot. The requirement to accurately establish the native location of something as small as a single protein molecule in a cell is a considerable one, and it is easy to go wrong in making such a judgment, particularly if small errors can lead to a different interpretation.

Once again, we confront the strong-microreductionist principle in how we go about sorting this out. Even something as technical as choosing the right exposure time may be based on assumptions about the native state—that is, where we expect the material to be found and where we do not expect it to be found—when these are presumably the facts that we are trying to ascertain. The strong-microreductionist viewpoint allows us the privilege of using our model, not merely to make predictions about what we may see, but to make measurements in a way that makes it more likely that we will see what we are predicting.

The Experiments

The Cytoplasm Once Again

As with the cell-fractionation experiments, the cytoplasm was the central issue in the autoradiography studies. In this case, grains found over the cytoplasm were assumed to be artifacts due to background noise, or outliers as the result of overexposure. The cytoplasm was usually excluded a priori as a natural repository for the labeled protein even when grains were found there. The same thinking was applied to other areas not thought to contain the protein, such as the extracellular space. If it could be assumed that the appearance of grains over these compartments was artifactual, then the task was to eliminate the artifact and correct the data set. This was done in a variety of ways. As mentioned, the exposure time could be adjusted to minimize the presence of grains over these areas; the grains could be mechanically subtracted as background noise; the tissue or tissue sections could be washed prior to the application of the emulsion to reduce the presence of such artifacts; or all of these methods might be employed.

In any event, the presence of labeled protein in the cytoplasm, extracellular spaces, and other places thought unlikely to be natural locations for the material was assumed to be artifactual at the outset, and the autoradiography experiments were designed and performed in a fashion to eliminate or exclude such artifactual sources. Consequently, the results were seen as validating the assumptions that gave rise to the particular experimental design and analysis a posteriori. And as such, the autoradiography results provided external validation for the cell-fractionation data. Applying appropriate assumptions, both sets of experiments led to the same conclusion. And however circular the reasoning, the common conclusion (vesicle passage and cytoplasmic exclusion) reinforced the assumptions that led to it.

Thus armed with parallel sets of assumptions, the two types of experiments led to the same conclusions. But it was all of a piece; mutually validating. Indeed, if the assumptions, whether procedural or interpretational, were warranted, it could be properly said that both experiments affirmed the vesicle theory. It could even be said that the autoradiography experiments truly outlined the dynamical features of the vesicle pathway. That is, radioactive protein actually moves through a series of vesicle compartments from the ER through the Golgi apparatus to the zymogen granules, finally ending up in the lumen of the duct system, as postulated. But without these assumptions, there was no clear evidence that any of this happened beyond synthesis on ribosomes, storage in the zymogen granules, and secretion into the duct lumen.

Gold Beads

Many variants of the cell-fractionation and autoradiography experiments were performed over the years in different systems by different groups of scientists. In most, though not all, similar assumptions were made, and not surprisingly, similar conclusions were drawn. Also, with time new methods were developed that provided far better resolution than was possible with autoradiography. Indeed, it even became feasible to identify the location of specific protein species within sections of cells with something approximating the needed resolution. This was achieved by chemically linking small opaque bodies, such as microscopic gold beads, to antibodies for particular proteins. These complexes were in turn bound to the protein of interest in sections of tissue. The protein's location was marked by the position of the opaque body. Although the dynamic

The Experiments

element was lost because the marking object had to be added to tissue sections, not living cells, the new method offered the possibility of identifying specific proteins, in particular small cellular compartments, down to the smallest objects in the vesicle model (the 40-nm-diameter microvesicles or the cisternae of the ER and Golgi). Now, it seemed, at least the location of the proteins might be clear.

But as significant an improvement as this was, the presumptions that had clouded the prior autoradiographic studies did not disappear. For example, common practice was to wash tissue samples before viewing them with the microscope in order to remove excess gold beads that might be inadvertently—that is, artifactually—attached to the tissue section, so as to better distinguish the signal from this noise. But as with autoradiography, washing could remove material from natural as well as artifactual sites, and not necessarily in equal proportions. How could one know what one had done, except by assumption?

But more important, when results did not support the vesicle theory—that is, when beads were not found where they were expected, or were found where they were not expected—ad hoc hypotheses were constructed to accommodate observation to theory. For example, in the first study of this type performed on the pancreas, the gold beads associated with a particular digestive enzyme were *not* found in the cisternal spaces of the ER. This seemed to falsify a central element of vesicle theory—the initial sequestration of digestive enzymes. But confidence in the vesicle model was so strong that this possibility was not entertained. It was assumed without evidence that the gold beads must have been present initially, but were lost during sample preparation.

When, in the same samples, protein-conjugated beads *were* found over the cytoplasm—that is, where the vesicle theory said they should not be—it was concluded that while the preparative procedures had caused the loss of material from the cisternal spaces of the ER, they had simultaneously added these proteins with their gold beads to the cytoplasm. *What was actually observed was thought to be a deception. What was inferred was real.* And what at first glance seemed to be powerful evidence *against* the vesicle theory—the absence of digestive enzymes in the cisternae of the ER and their presence in the cytoplasm—was seen as evidence supporting the model.

In a later study, quite different results were reported. These investigators found gold beads conjugated to digestive enzymes in the cisternal spaces of the ER, as well as in the cytoplasm. But following the same set of assumptions, they saw the beads in the ER as convincing proof of the sequestration hypothesis, while their presence in the cytoplasm was assumed to be artifactual. Moreover, when one looked carefully at the distribution of the gold beads in and around the ER, even the conclusion that sequestration had been demonstrated was not clear. Most of them could not be unambiguously attributed to the cisternae. They were near them all right, but not obviously within them. In fact, the great majority seemed to be over the membrane enclosing this space, associated with attached ribosomes, or over the cytoplasm just outside the reticulum. Thus, in the end, the advances made possible by the introduction of new immunocytochemical techniques, using gold beads and the like, furnished no protection against the power of prior expectations.

The Illusion of Dynamics

Why all this complexity and confusion? And why the myriad assumptions? After all, radiolabeled amino acids are incorporated into new proteins, and in the normal course of events these proteins travel through the cell along their natural path or paths. Were we not simply attempting to follow this process? As we have discussed, neither cell fractionation nor autoradiography are actually able to follow the events of transport *passively* or *disinterestedly*. They do not merely allow the dynamics to unfold before our eyes. The experiments were *proactive*, to use today's argot. The system had to be altered, and nontrivially so, to make the measurements. Whether the cell was broken open by homogenization or the tissue was prepared for microscopic viewing by various chemical and physical manipulations, the natural state was severely disrupted.

Instead of the real dynamics of the process, an ersatz dynamics was measured that gave only the illusion of the real thing. However secure the vesicle model seemed—and it seemed secure to most workers—the observations crucially and inescapably did not demonstrate the dynamics of transport in a fashion that was independent of the model's truth or validity. Only if we adjudged it to be true at the outset did the ersatz become real. But doing this turns the process of research on its head. Rather than trying to establish the goodness of fit of a theory to observation, the experiments sought to fit observation to theory. The question was not whether the theory was consistent with the observations, but whether the observations were consistent with the theory.

All of this was attributable to the strong-microreductionist position. In its terms, it was not significant that the experiments' meaning was dependent on the validity of the model. The model was not merely a device for learning, but had become an essential part of the description of actual events.

CHAPTER 11

To Be Parallel or Nonparallel

Before a theory explaining a certain process can be tested, the process must first be known.

Albert Einstein

Whatever the uncertainties, the list of experimental evidence consistent with the vesicle model became long and imposing. From the various evocative structures and forms seen with the electron microscope, one could easily imagine how things might occur as envisioned in the model. Not only that, but however circular the reasoning, those structures and forms were "predicted" by the model and were seen. As for the cell-fractionation and autoradiography experiments, though alternative explanations could not be rigorously excluded, the data provided a range of affirmative evidence that was consistent with the vesicle model, and given the reasonableness of the assumptions that were made, proved its applicability.

All in all, many things pointed to the vesicle model's correctness, even without examining its goodness of fit to the natural process. But how much faith could one really have in observations that, however affirming, did not involve testing the theory against the various properties of the natural process? The answer is not much. No matter how much evidence there was, no matter how reassuring because it fit so well with common preconceptions, if affirmations, not tests, represent the limit of our knowledge, we cannot even trust in a theory's cogency, much less its correctness.

The Case of the Two Williams

To see that this is so, try to develop a list of affirmative observations for theories that you know in advance to be false. It is easy. Such affirmations, however false and misleading, can be very convincing, very seductive. The simple fact is that no list of affirmative observations can prove a theory correct. Whereas a *single* negative observation, one failed test, can prove it false. To understand how this works, let us look at the case that led to the U.S. government's decision in 1905 to use fingerprints to identify criminals. At the time, the state of the art in the identification of criminals was a method developed by the French criminologist Alphonse Bertillon. In the Bertillon method, various physical features of the subject were measured, such as height, arm length, finger length, and ear width. From them, a composite profile was established.

In 1903 this method of identification, as well as the other means used by the government to identify felons, ran into a serious problem at the federal penitentiary at Leavenworth, Kansas. A guard was processing a newly convicted criminal named Will West. Will looked familiar, and the guard asked him if he had ever been an inmate at Leavenworth before. He said that he had not, but the guard remained suspicious and searched the records to check his hunch. He found that not only had a William West been an inmate at Leavenworth, but much to his surprise was currently incarcerated!

To Be Parallel or Nonparallel

William West's file contained the standard photograph, and it looked exactly like the person he was processing. To make sure that he was not dealing with two men who looked alike and shared the same name, the guard compared their Bertillon measurements. For all intents and purposes they were identical, inch for inch, centimeter for centimeter. And if that were not enough, the two men had uncommonly similar backgrounds. Both were born in Texas, left home at age 12 or 13, and had most recently resided in the Oklahoma territory. Both of their mothers were dead, and both mothers were also said to have been born in Texas. Finally, both men were bank robbers. Based on these facts—name, appearance, the Bertillon measurements, and history—the guard was confident that Will and William were one and the same person. Although he did not understand what Will/William was up to, he was sure it was no good. And yet, against all of this evidence, the new convict continued to deny that he had ever been to Leavenworth.

Fingerprinting had recently been introduced at the prison on an experimental basis. William West's file contained his fingerprints, and the guard decided to compare them to those he had just taken of the newly incarcerated Will West. Much to his surprise, they did not match. In spite of what appeared to be overwhelming affirmative evidence that Will and William were the same person, this one piece of data argued otherwise.

What evidence to accept? The long list affirming the identity of the two men, or this single piece of data arguing against it? The fingerprint was a new method that had not yet been fully tested. Most likely it was in error. If not, the guard would have to conclude that all of the other evidence was wrong, and he had no basis to assume that *any* of it was wrong. But the fingerprint pattern was reproducible: the same pattern could be obtained over and over again from the same individual and could be distinguished from what appeared to be equally reliable measurements on others. There seemed no basis for assuming that *it* was incorrect. The prison guard had been confronted with two conflicting sets of data, each of which seemed to be valid in its own right but led to diametrically opposite conclusions.

As it turned out, the problem was very easy to solve. The guard just looked for William West in the prison population. There were indeed two men, dead ringers for each other. The fingerprint had been correct, and all of the other evidence, physical and historical, though accurate enough, had been misleading. All the similarities between the two men that affirmed their identity collapsed like a house of cards when confronted by this one piece of negative data.

So, if our goal is to obtain scientific proof, it is not enough to accumulate evidence to support a particular theory or model, even if that evidence is extensive and appears compelling. If we place our trust in the simple enumeration of affirmative evidence, we can easily mislead ourselves, much as with Will and William. This is not to say that we should ignore affirmations of a model or theory. Of course not. After all, they may well be pointing us in the right direction. But it is just not enough, if scientific proof is what we seek.

To Test a Scientific Theory

While it must always remain conditional, affirmative evidence can gain meaningful authority in only one circumstance: when it is obtained in the context of an experiment in which a negative outcome is also a possibility, that is, when it is collected as part of a scientific test in which the theory has the potential of being rejected by the evidence rather than affirmed. With the possibility of rejection, affirmative observations gain the power of discrimination. But it is important to remember that such authority can be gained only in the shadow of failure, when a negative outcome is possible.

We have already considered the first requirement for the scientific exploration of a theory or model—we must compare the properties predicted by our model to those of the intact functioning system. We now have the second requirement. One outcome of our experiment must be that our theory or model is shown to be incorrect or false. As obvious as this may seem, it is not uncommon in science for theories to be studied, even extensively explored, without ever being put to the test. Sometimes it is impossible for one technical reason or another. And some theories are so vague or so complex, or both, that they are simply not amenable to scientific testing. Psychoanalytic theory is a good example.

To understand that there is not some absolute character or quality of evidence that is important here, but merely the notion of falsification, imagine the scenario with Will and William being reversed in the following sense. What if they displayed identical fingerprints along with all of the other evidence of their identity? Would we now be certain that we were dealing with one and the same person, whatever trick he might be playing? But what if the guard then brought the two men into the same room? Now the fingerprint that had seemed to be such powerful evidence would become as suspect as all the rest of the data. Seeing, touching, and talking to the two men would become the commanding data, falsifying the hypothesis of identity and outweighing what would now be the very questionable fingerprint.

The philosopher of science Karl Popper argued for the importance of falsification and deductive testing, as well as the need to appreciate the contingent nature of all scientific knowledge. Popper said that our most secure knowledge of the natural world is what we know to be *incorrect*, however uncomfortable this may be for the scientist who has spent a career trying to gain an affirmative understanding of nature. Conversely, Popper maintained that evidence affirming a theory or model, even if the result of a positive test, must always remain conditional, because the possibility perpetually exists that some future piece of data will show the theory to be false.

In the end though, Popper cautions against too rigid an adherence to the principle of falsification. He says that as a matter of *practice* negative evidence should not always be accepted as proof of the failure of a theory. Popper asks why one should throw away a good theory simply because of one negative result. But Popper was flummoxed about how many negative results would be required to reject a theory. Two, twenty, two hundred? How could one say? There seemed no logical, rigorous end point, no decisive criterion, if one allowed that a single negative test did not necessarily falsify a theory. Likewise, was it really true that positive evidence could never be conclusive? For example, the presence of two Williams in the same room seems pretty conclusive evidence for the hypothesis that there are two men.

Such ambiguities and uncertainties about our theories and models probably can never be overcome solely as a matter of logic, but the critical point remains. Unless we confront our hypotheses with the possibility of their falsification, we run the risk of deluding ourselves about the state of our knowledge. And such delusions open the door to unsubstantiated and hence dogmatic belief. Without applying potentially falsifying tests, even a most impressive list of affirmative correspondences may in fact teach us little about the real nature of things, though we may certainly beguile ourselves into thinking that our understanding is substantial.

The Phenomenon of Nonparallel Secretion

Critical tests of the vesicle model, though rare, were not totally absent. There were some instances in which the model was tested against the properties of the secretion process in the intact cell. As it happened, many of these studies were carried out in my laboratory, and in the remainder of this chapter and the next I will tell how, along with my students and associates, I came to question the well-regarded and widely accepted vesicle theory.

There was an old, mostly forgotten dispute between Ivan Pavlov and one of his students, B. P. Babkin. Pavlov was among the first to understand that the body needs to regulate or control its biochemical reactions in one way or another for them to be useful. It was with this understanding in mind that at the turn of the twentieth century he proposed that the various chemical reactions of digestion were regulated. Pavlov believed that this was accomplished by the action of the nervous system. In response to digestive need, the rates of secretion of the various digestive enzymes produced by the pancreas would be adjusted independently of each other in response to signals from the brain.

Though proof for his hypothesis was minimal, he thought that it made good sense nonetheless. If digestion was regulated chemically, then varying the rates of secretion of different enzymes would be the reasonable way to accomplish this goal. And the nervous system seemed the only means available to produce the necessary variations (hormones had not yet been discovered). Why, Pavlov thought, would the body secrete an enzyme that splits starch (such as amylase) when protein is in the bowel to be digested or secrete enzymes that break down protein (such as trypsinogen or chymotrypsinogen), when a starchy meal awaits digestion? As he imagined it, the mixture of enzymes secreted by the gland would be altered to accommodate the particular meal undergoing digestion.

However sensible his idea, Pavlov did not pursue it in any depth experimentally. In fact, not long after making this proposal he gave up his highly successful research on the gastrointestinal system and digestion altogether for reasons that are a story unto themselves. And by the time of the October revolution of 1917 he had become totally immersed in studying the conditioned reflex. Because of his worldwide reputation, as well as the political utility of his ideas on conditioning to the communist state, Pavlov remained a powerful figure in the Soviet Union until his death in 1937.

His assistant, Babkin, on the other hand, left Russia and moved to Canada, where he joined the faculty at McGill University in Montreal. It was at McGill, some 30 years after Pavlov's original proposal, that Babkin tested his mentor's hypothesis about digestive enzyme secretion. And when he did, he decided that Pavlov had gotten it wrong. The digestive enzymes were not secreted in varying proportions to meet the needs of particular meals; to the contrary, they were released at constant or set ratios under all conditions, regardless of the nature of the meal undergoing digestion. If the secretion of one enzyme was increased, so was the secretion of all others, and to the same proportionate degree. Babkin called this *parallel secretion* because the various proteins responded to stimuli in parallel with one another. In Babkin's view, there was no chemical regulation of digestion. All of the various enzymes were secreted into the intestinal tract together, en masse, ready to deal with whatever food had been ingested. The only way that the body could regulate these reactions was by varying the total *amount* of enzymes secreted into the intestinal tract in some proportional relationship to the size of the meal.

To Be Parallel or Nonparallel

Although Babkin's conclusion was clear enough, his evidence was less so. But whatever the evidence, perhaps because he seemed to be proposing less rather than more—or perhaps because Pavlov was no longer interested in the gastrointestinal system, or had become scientifically isolated in the Soviet Union—Babkin's view prevailed with little if any apparent resistance. One consequence of this was that it became common practice to follow only one enzyme (usually amylase) when measuring the secretion of digestive enzymes. If they all behaved the same, then what purpose would be served by measuring more than one?

I became interested in this old dispute in 1964 soon after finishing my PhD degree at the University of Pennsylvania. I was preparing to move to Boston to take my first academic job at Harvard and thought that in the few months available before I moved it might be interesting to revisit the dispute between Pavlov and Babkin, using the experimental system I had invented for my thesis work. It was an organ culture preparation of the pancreas in which secretion from the duct system of whole glands could be studied in a controlled in vitro setting. By assessing enzyme secretion under these conditions, and by using new and improved methods for measuring their presence, I hoped to be able to make a more conclusive determination about parallel secretion than had either Babkin or Pavlov. Toward that end I compared the secretion of *two* closely related digestive enzymes, trypsinogen and chymotrypsinogen. Both are proteolytic enzymes (enzymes that degrade proteins by splitting their amino acid chains); they are approximately the same shape and size and share fully half their amino acid sequences. But as similar as they are structurally, they are functionally distinct. Trypsinogen cleaves peptide bonds (the bonds that hold amino acids chains [peptides] together) at linkages that contain basic amino acids, whereas chymotrypsinogen is effective at aromatic amino acid sites.

These simple experiments set me on a path that has been strewn with equal measures of exciting discovery and contentious controversy about how proteins are transported through and out of cells. What I found was that a particular peptide hormone that stimulates the secretion of digestive enzymes by the pancreas favored trypsinogen secretion over chymotrypsinogen secretion. Although the hormone had enhanced the secretion of both enzymes, if one looked at the data closely it was clear that this increase had occurred not in parallel, but differentially. I called this effect *nonparallel secretion* to contrast it with Babkin's *parallel secretion*.

The observation raised two consequential issues. First, if the proportions of different digestive enzymes in secretion could vary, then Pavlov's hypothesis about the regulation of digestive reactions might be correct, and the question of regulation needed to be revisited. Not long after this discovery, my students and I began such a reconsideration. In a variety of studies we were able to show that the chemical reactions of digestion could be regulated by varying the rates of secretion of different enzymes, just as Pavlov had proposed.

The second point concerned the mechanism of secretion. What, if anything, did this property tell us about the mechanism of secretion? To this point I have discussed the different proteins secreted by the exocrine pancreas as if they were all one and the same, all transported to and stored in the zymogen granule in exactly the same manner, and subsequently secreted in an undifferentiated fashion, much as Babkin had envisioned. But if they could move differentially, perhaps even independently of each other, then this convenient generalization was clearly ill suited to the actual situation.

To Be Parallel or Nonparallel

And if this was so, then what about the vesicle model and its proposed mechanism of secretion, exocytosis? Exocytosis was thought to be a random event; the consequence of chance chemical interactions between a site on the membrane of a given secretion granule and a receptor site on the cell membrane. As discussed earlier, such interactions were thought to cause a fusion of granule and cell membrane, which in turn would lead to the formation of a congruent hole in both membranes, through which the product would exit the granule into the extracellular environment. But however the granules interacted with the cell membrane and however product was released from them, the central point was that they were chosen for expulsion *randomly*—that is, in an accidental fashion.

In this case, the enzymes could only be secreted *in parallel* with each other—that is, in fixed or invariant proportions, as Babkin argued. This would be true even if they were stored in different granules. If exocytosis occurred randomly, there would be no way to distinguish among the granules, whatever their contents. In this circumstance, the composition of secretion would be unchanging and unchangeable as long as the contents of the source pool—that is, the granules—also remained unchanged.

It was in this way that parallel secretion and the vesicle model were joined. Indeed, parallel secretion was a *prediction* of the vesicle model. If the model was correct, then secretion should be parallel at all times, under all circumstances. If, on the other hand, variations in composition not attributable to changes in the source pool were seen, we would have to conclude that the vesicle model was false, at least as the sole mechanism of secretion. And since this was in fact the case, there seemed to be another mechanism of secretion instead of or in addition to the one proposed in the vesicle model.

But why not keep the exocytosis model and simply conclude that the release of granules occurs selectively, not randomly? Perhaps there are divers granule types that contain different enzymes or groups of enzymes that can be selected among by the exocytosis mechanism. Isn't this a possibility? Yes, it is, if distinct granule types and selective exocytosis exist. However, if either does not, the vesicle model would have to be rejected as the sole mechanism of secretion and perhaps altogether. As said, if there were different granule types, but exocytosis could not distinguish among them, secretion would be parallel. Alternatively, and trivially, if selection for exocytosis could occur, but divers granule types did not exist, secretion would be parallel. Only if both occurred could the theory be saved, though it would still have to be proven true.

But such a mechanism of secretion would be far more complex than the already extremely complex one originally envisioned. And as it happened, there was no evidence either of dissimilar granule types or selective exocytosis, and the common wisdom was that neither existed. This was principally a matter of parsimony. A mechanism of random or nonselective exocytosis made fewer demands—required fewer propositions—than a selective version of the model, and was favored in the absence of evidence to the contrary. Many years later, using the immunocytochemical methods that I described in the preceding chapter, in which gold beads are attached to antibodies for specific enzymes, it was demonstrated that all of the digestive enzymes, or at least all of the numerous enzymes that were measured, were found in each and every granule, in each and every cell.

Thus, it was found that enzyme-specific granules do not exist in the acinar cells of the pancreas. Since the evidence seemed to argue strongly against intergranular order, it argued as well against exocytosis being able to explain nonparallel secretion. We might still be able to save the exocytosis model if we allowed for the possibility of even greater complexity. For example, perhaps all of the enzymes were stored together in single granules all right, but the *proportions* among them varied in some particular ordered fashion. Conceivably, such variations, could, if they existed, provide different granule types that could explain nonparallel secretion with an exocytosis mechanism. But this was an awkward idea. How many types of granules would there be—two, ten, hundreds, thousands? And, most important, on what *functional* basis would the different proportions be determined? What would be the utility of this kind of partitioning? A little more of this, a little less of that, to what purpose? And to accommodate such a mechanism we would have to propose that, seemingly without clear functional purpose, a complex system had evolved to segregate products this way and to select particular granules for secretion.

Cyclic AMP and Exocytosis

However likely or unlikely this may seem, something else argued strongly in favor of random release if it occurred as a result of exocytosis. The means of mediating cellular responses to stimuli—that is, of coupling stimulus to response—seemed to exclude any other possibility. It was discovered that a unique chemical substance called a second messenger, a cyclized nucleotide called *cyclic adenosine monophosphate* (cAMP), was manufactured by cells in response to external stimuli (the first messenger being the external stimulant itself). It was this second substance that was thought to be the proximate cause of the response.

It did not matter how one caused the cell to secrete, what chemical initiated the response, whether it was a nerve to the secretion cell or a hormone. In either case, cAMP production would be elevated. To the cell it was irrelevant what caused the production of cAMP. All it would know was that it was produced. Eventually a rather complex sequence of events was identified that was interposed between the first messenger and the response (third messengers and so forth), but this did not seem to change anything. The cellular response would always be the same, unavoidably stereotyped and qualitatively invariant, as long as all stimuli acted through at least one common intermediate.

In this view, secretion was just like the contraction of muscle. Whereas a muscle can contract with greater or lesser force, a gland can secrete more or less material. However one goes about eliciting a response, there is only one —contraction or secretion—that can be varied only quantitatively. Thus, second messenger theory argued persuasively against selective exocytosis however one might imagine it occurring.

This was the bottom line when I published my first observations on the phenomenon in *Nature* in 1967. Since there was no evidence of selection either in granule contents or granule release, the community of knowledgeable scholars believed that random vesicle mechanisms were fully responsible for secretion. And in addition stimulus-response coupling seemed to require a single stereotypic response to any and all stimuli if vesicle mechanisms were responsible for secretion. Given this, what happened subsequently probably should not have been surprising, though it certainly surprised me.

My findings, and all subsequent reports of nonparallel secretion, from my laboratory as well as from others—whatever the species, the animal model, the methods of measure-

To Be Parallel or Nonparallel

ment, and the modes of stimulation—were commonly labeled *artifacts*. To many there seemed to be no choice. It was argued that if the measurements had been properly made, secretion would have been found to be parallel. Some workers reported that in their hands it was found to be parallel—and not surprisingly, they maintained that theirs were careful hands. They believed fallaciously that if they could find constant proportions in one particular situation or another, this would mean that it could be expected in *all* situations. Only after some 15 years of controversy, and not before some 50 examples of nonparallel secretion had been peer reviewed and published, was disbelief overcome—though no doubt there are still some who are skeptical in their heart of hearts.

Well then, was the evidence of nonparallel secretion, after 15 years of controversy, now seen as falsifying the vesicle model, especially given the substantial weaknesses in the evidence for that model at the time? The answer was a resounding no; such a conclusion was not seriously considered by most experts. When it became clear that nonparallel secretion was not an artifact, but a real phenomenon, it was assumed that somehow, in some way, it was consistent with exocytosis and the vesicle model. Indeed, nonparallel secretion was seen as providing new information about the vesicle mechanism. The observations had not falsified the theory at all, but clarified and refined our understanding of it. Nonparallel secretion was *proof* that the selective exocytosis of distinct granule types occurred, however unlikely it may have seemed heretofore and whatever the lack of direct evidence for such events had been.

This circular turn of mind should be familiar by now. Remember our discussion of the omega figure? An observation that suggests the possibility (the hypothesis) of a particular mechanism can be seen as evidence for that mechanism.

The central difference between the omega figure and nonparallel secretion was that in the latter case, the phenomenon was clearly a property of the natural secretion process. But this made the inference no more reasonable. The observation of nonparallel secretion no more provided evidence of selective exocytosis and distinct granule types than the omega figure provided evidence of exocytosis.

Assumptions

Still, if selective exocytosis was the only show in town—the only way that secretion could occur—it had to explain nonparallel secretion somehow. But there was another way, an alternative. It was grounded in established physical law and envisioned secretion in a totally different manner. It could explain nonparallel secretion, as well as a variety of other observations that had been problematic for the vesicle model, in a relatively straightforward way. However, the alternative hypothesis had never been tested experimentally, either in its own right or in comparison to the vesicle model. As a result, there was no evidence to offer in its behalf.

This was neither an accident nor an oversight. If nonparallel secretion was a controversial observation, then the mechanistic alternative to the vesicle theory was controversial squared or cubed! To most cell biologists, it was not merely implausible, but most certainly impossible, and no purpose would be served by testing a hypothesis that was known in advance to be impossible. There were two imposing reasons for drawing this conclusion. The first was based on foundational understandings of the physical and chemical nature of the cell and its protein contents. The second seemed even more inescapable. It could not be imagined how the living cell could survive its occurrence.

These were very compelling reasons indeed, and they were at least in part responsible for the abundant confidence in the vesicle theory and the willingness of investigators to make the sorts of assumptions we have been discussing. They were also important reasons why it was thought that an understanding of the mechanism could be achieved by means of the strong-microreductionist approach. However compelling, I found myself disagreeing with both reasons. I did not think that the alternative contradicted any known—that is, experimentally or theoretically established—understanding of the physics and chemistry of cells or their protein molecules, and I could easily imagine how cells could survive its presence. My dissent was rooted in the realization that the reasons for concluding that the alternative model was impossible were themselves unproven suppositions, though they seemed axiomatic at the time.

The Membrane Transport of Proteins

Well, what exactly was this unlikely alternative mechanism? Its consideration brings us back to the question we discussed in the last chapter, about the presence of digestive enzymes in the cytoplasm. If proteins that are to be secreted are found in the cytoplasm of the cell, freely diffusible, then they might be released directly across the cell's enclosing membrane, molecule by molecule, as a result of their own movement, instead of or in addition to being released by vesicle processes. This was the impossible alternative—the *direct transport* of protein molecules across biological membranes.

At the time, and indeed until relatively recently, it was the opinion of most knowledgeable scientists that protein molecules could not cross biological membranes by interacting individually with them and passing through according to their own properties and concentrations, as do small atoms and molecules such as sodium, potassium, glucose, various amino acids, and so forth. That is to say, the physical properties and laws that determine the transport of other substances across membranes were not thought to apply to protein molecules as a class. It was understood that proteins could not cross membranes as the result of their *diffusion*—the inherent propensity of molecules to move due to their thermal agitation or kinetic energy. This was the basis for the pivotal assumption that the cytoplasm was free of all proteins secreted by the cell. If they were present, they would languish there, unable to leave.

Notwithstanding this belief, if membrane transport mechanisms *did* exist, then the fact that the composition of secretion could vary—that secretion could be nonparallel—could be explained quite simply. It would be the consequence of known and quantifiable differences in the physical properties of the transported molecules themselves, those properties that determine the facility with which they diffuse in liquids, like their shape and size, as well as differences in their concentration. Indeed, simple physical diffusion was the parsimonious hypothesis for the movement of all molecules across short distances whatever the nature of the barrier, including biological membranes. It was against this standard that more complex notions of transport, even ones far simpler than the vesicle model, had to be judged. The proper and traditional question was whether an explanation based on diffusion was sufficient. Could it account for the data, or was a more complex model necessary?

Often simple diffusion alone could not account for the permeability of a biomembrane to a particular substance. More complex mechanisms were needed. But this realization did not open the floodgates to any and all conceivable theories of transport. Scientific models remain tethered to the simplest hypothesis, even when that hypothesis is found wanting. The scientist asks what *minimal* modification must be made to this simplest model for it to be compatible with nature. The goal is to avoid unnecessary complexity by making only such additions to our constructs as nature absolutely requires. This is the path of parsimony and rigorous exploration. Otherwise, there is little to constrain our imagination from conceiving possibilities of inordinate complexity. As we have discussed, the vesicle theory did not emerge from such a process of gradualism, step by halting step, as required by the accumulating evidence. It emerged essentially fully formed, a homunculus of the human imagination based on the appearance of objects within the cell, bolstered greatly by the assumption we are now discussing. The possibility of the diffusion of protein molecules across membranes is nowhere to be found in the genesis of the vesicle theory, except by its a priori exclusion.

In addition to providing different explanations for the movement of molecules across biological membranes, vesicle and diffusional models are proper dialectical alternatives. Either proteins—or any other molecules for that matter—move within cells and cross relevant membranes as a consequence of their own inherent motion, whether by simple diffusion or more complex chemically mediated diffusional mechanisms, or they are carried passively by an external mechanism of some sort, such as in vesicles. This is the stark difference between the two conceptualizations, and it offered the opportunity to design experiments to distinguish between them. Such comparisons would be very powerful and very informative, because they could include or exclude one or the other of the possibilities, thereby allowing us to limit our investigation.

I should be clear that a diffusional model does not presuppose the exact nature of the transport mechanism, any more than a vesicle model delineates all the details of the mechanism it proposes. As I have said, simple physical diffusion might be sufficient, or more complex mechanisms might be required. But in either case, it would be the properties of the transported substance that provides the driving force for movement. Specialized membrane-embedded, chemically mediated transport systems might well exist, or perhaps both simple and complex mechanisms might be found side by side in the same membrane. Whatever the particular incarnations, one could make predictions about their expected properties to compare to those predicted for vesicle-mediated transport, testing both against nature.

Making such comparisons was thought to be pointless by most workers—an exercise in futility, if not stupidity. Membrane transport mechanisms were quite impossible for protein molecules, whatever the molecule, whatever the membrane, and however one envisioned passage occurring. Though this was admittedly the parsimonious model, and known to account for the movement of many small inorganic and organic substances across membranes, it was confidently understood that it did not apply to proteins, including the digestive enzymes that I was studying. It was for this reason that *external* mechanisms, such as vesicles, were needed. Indeed, it was their raison d'être. Without them transport could not occur, and of course it did. And turning things around in that familiar circular way, didn't evidence suggestive of vesicle mechanisms argue against a membrane permeability to protein molecules? Why else would vesicle mechanisms exist?

This zeitgeist was grounded in the two axioms already mentioned. First, our basic understanding of the structure of the biological membrane and of protein molecules made their direct passage across biomembranes impossible. And second, the cell could not survive the existence of such a process, because crucial cellular proteins would be lost and ion gradients and the like dissipated. In general, their presence would wreak havoc on the carefully constructed cellular steady state.

Even though at the time relatively little was known with certainty about the structure of biological membranes, a particular view was widely accepted among biologists. In this view the membrane was a nonpolar structure comprised mostly, if not wholly, of lipid molecules. What evidence there was suggested that the lipids were organized in a bilayer—that is, as a layer two molecules thick. It was this slight layer of lipid that kept the cell separate from its watery environment. The cell membrane was life's oil to the environment's water.

This perception was supported by a variety of facts. Most important, polar substances—substances soluble in water but not oil—crossed biological membranes poorly. In the language of the field, their diffusion was highly restricted. Whereas substances soluble in oil, nonpolar substances, of all sorts and sizes, such as various animal and plant fats, crossed readily. This was realized in the early years of the twentieth century when the permeability of various membranes to a variety of polar and nonpolar molecules was first measured.

To the degree that polar molecules crossed membranes at all, their passage was highly correlated to size. For example, although highly restricted in comparison to their unfettered movement in a beaker of water, very small polar compounds such as urea crossed relatively easily, as did small atoms such as the potassium ion. As the size of the substance became larger, its ability to penetrate became increasingly, indeed exponentially, diminished. Notably, when polar hydrocarbon molecules were examined, increasing the length of the carbon chain from one to two to three carbons and so on greatly limited passage. By the time one reached molecules of, say, the size of mannitol, an alcoholic sugar that is six carbons long, permeability was reduced by several orders of magnitude.

The popular model that sought to explain these characteristics suggested that biological membranes were dotted with and traversed by small water-filled channels. It was only by diffusion through these *pores* that polar substances could cross the membrane. As the size of the diffusing molecule approached that of the pore, passage would become increasingly restricted. The percentage of molecules *reflected*—that bounced off the edges of the pore back into the source compartment—would increase greatly. And obviously, a molecule that is larger than a pore cannot enter it. The membrane would be absolutely impermeable to such substances.

How did protein molecules fit into this picture? Could they cross the lipid layer or through the small pores thought to be embedded in the membrane? It turns out that most proteins are polar molecules. Their surfaces are studded with numerous charged (polar) groups that can share electrons with water, and hence they can dissolve in it. But they seemed too large by far to pass through water-filled pores of the size projected for biomembranes. Proteins were among the largest known biomolecules, comprised of chains of amino acid hydrocarbon subunits *hundreds* of units long. The average protein molecule was 2000 to 3000 times the mass of a molecule of water and some 20 to 100 times it diameter; it was some 200 to 300 times the mass of the relatively impermeable mannitol and several times its diameter. All in all, the water-filled pores seemed far too small to accommodate huge protein molecules.

To Be Parallel or Nonparallel

Furthermore, the polar groups on the surface of proteins kept their solubility in oil low. Given a choice between water and oil (the fat of the membrane), proteins would invariably choose water. Thus, they presumably would not enter or cross a biological membrane by diffusion through its lipid substance. If proteins could not pass through pores in the membrane or cross the lipid barrier, then there seemed to be no way for them to cross due to their own motion. As the physical and chemical properties of membranes and proteins were understood at the time, the possibility of a membrane permeable to protein seemed out of the question.

This made good biological sense. It even seemed necessary. This brings us to the second axiom—that the cell could not long survive such a permeability. After all, if proteins are the critical substances of life, does it not make sense that the cell membrane evolved to prevent their loss? It seemed that a permeability to protein molecules would be incompatible with life, or at least a threat to be avoided if at all possible. Wouldn't it allow crucial life-sustaining molecules to simply leak out and be lost? Didn't it make far more sense for important intracellular proteins to be constrained by the membrane to prevent their loss, and that special mechanisms like those proposed in the vesicle theory had evolved to overcome this barrier in order to allow for the release of a subclass of proteins that work in the cell's surroundings? A membrane that is impermeable to protein molecules in conjunction with a vesicle mechanism that allows particular proteins to bypass that barrier seemed exactly what Mother Nature had ordered.

And yet it was curious. In spite of the widely held belief that membranes were impermeable to proteins, one never came across this justification for the vesicle theory in the literature. If it was the bedrock, then why not just say so, and say so explicitly? "Vesicle mechanisms must be involved because the only alternative, the direct membrane transport of proteins, is absolutely impossible." The reason for the silence was that there was no experimental evidence that membranes actually presented an absolute barrier to the digestive enzymes or any other protein molecule. There were just strongly held beliefs.

CHAPTER 12

The Tests

I learned to believe nothing too certainly of which I had only been convinced by example and custom.

René Descartes

And so nonparallel secretion had raised the radical possibility that membranes might be permeable to protein molecules. While the phenomenon could be explained in terms of the vesicle model, the evidence, as well as notions of parsimony, seemed to argue against it. It was with this in mind that my laboratory began a series of experiments designed to test the predictive power of the vesicle theory against the observable properties of nature.

The main focus of this work was on evaluating the expectation that the membranes of the acinar cell were impermeable to its secreted proteins. If this was correct, as was widely believed, it would be powerful evidence for the vesicle theory. On the other hand, if it was not—if the relevant membranes were permeable—then, we believed, the whole hierarchy of assumptions that had bolstered the vesicle model would come crashing to the ground, and the unmistakable weaknesses in the evidence for this model would be disclosed.

At the time this was a perilous undertaking. Doubting what was understood to be true, and contemplating that the impossible might occur, was not likely to engender great support. In any event, as a young scientist my student and I plunged into this effluvium on what was to be a lonely and unpredictable voyage, with not much more in mind than that scientific theories should be tested and the law of parsimony obeyed.

The experiments reached their culmination in the revolutionary determination that particular membranes of the acinar cell were permeable to digestive enzymes. This not only brought the notion that vesicle mechanisms provide the sole means of protein transport into doubt, but raised larger questions about cellular mechanism and organization. Perhaps the most important was that it made us wonder whether a general prohibition against the permeability of biological membranes to protein molecules was justified.

Though the experiments were diverse, and carried out over a period of some 30 years, they almost exclusively concerned two related predictions of the vesicle theory. The first we have already discussed. It is that if the vesicle theory is correct, the digestive enzymes should be ex-cluded from the cytoplasm of the acinar cell. Whereas if secretion occurs as a consequence of diffusion-based membrane transport, the cytoplasm should contain these substances. In this case, the cytoplasm would serve as source and repository, interposed between the duct system and cellular storage sites such as the zymogen granule. If we could determine whether digestive enzymes were in the cytoplasm or excluded from it, we would be able to clearly distinguish between the two models. A nonvesicular membrane transport mechanism required their presence; while their absence would affirm the vesicle pathway as the exclusive means of transport.

The Tests
The Cytoplasm Redux

In Chap. 10 we discussed two approaches to making such a determination, as well as the interpretative problems they faced. For example, there seemed to be formidable and seemingly refractory difficulties in judging whether digestive enzymes in high-speed supernatant fractions of homogenized tissue were normal cytoplasmic constituents or merely artifacts—although, as explained, they were widely assumed to be the latter.

In the 1960s and early 1970s we tried to solve this problem in several ways. In one we compared fractions from tissue in *different* specified functional states, not just using a single state or ignoring state altogether, as had been the practice. If the presence of digestive enzymes in the supernatant fraction was due solely to artifactual causes, then the percentage there should be constant whatever its value—10, 20, or 50 percent of the total—regardless of the prior physiological state of the tissue. That is, if the methods of homogenization, separation, and measurement were the same for all samples, and if it was these procedures and these procedures alone that introduced digestive enzymes into the supernatant fraction, then the amount found there should not vary—or, more accurately, should vary only randomly—whatever the prior state of the tissue.

On the other hand, if the presence of digestive enzymes was *not* entirely artifactual, if there was a physiological component, then differences might be seen in material recovered from functionally dissimilar glands. Such differences, whatever their particular nature, would confirm that digestive enzymes were natural constituents of the cytoplasm. In addition, we expected that if differences were found they would be distinct for particular enzymes. For example, if nonparallel secretion was the consequence of the preferential release of a certain enzyme from the cytoplasm relative to others, we might find less of it remaining in the supernatant fraction than another less favored enzyme. That is, we might find that the *proportions* of various enzymes had changed in the cytoplasmic fraction as well as in secretion.

We carried out such an experiment by comparing tissue from glands in an unstimulated state to tissue from actively secreting glands, using the same stimulant that had produced the original nonparallel response (the hormone cholecystokinin-pancreozymin [CCK]). We hoped to determine whether the digestive enzyme content of the supernatant fraction had been altered in a fashion that reflected the nonparallel secretion that had taken place. Remember that this hormone enhanced trypsinogen secretion to a greater extent than chymotrypsinogen secretion. If this was a consequence of their differential release from the cytoplasm, then the cytoplasm-containing cell fraction might mirror what had occurred in secretion by showing a selective depletion of trypsinogen.

This is exactly what we saw. The supernatant fraction was depleted of the enzyme whose secretion had been favored. Unless this change could be attributed to other sources, other fractions, with the supernatant just parroting them artifactually—which it turned out it could not—this would prove the existence of a cytoplasmic pool of digestive enzyme and confirm the phenomenon of nonparallel secretion.

From the perspective of this book, the crucial feature of this experiment was the use of physiological state as a means of comparison. The observations had gained meaning by being anchored in events that took place in the whole cell prior to its homogenization. This had infused the microreductionist approach of homogenization and separation with what it lacked: the information needed to distinguish among functional hypotheses. Because the various artifacts of homogenization could be assumed to be identical for all samples regardless of the tissue's prior state, whatever differences remained could be attributed to properties of the intact object.

The Cytoplasm in Situ

The presence of a cytoplasmic pool was also explored in a variety of experiments in whole functioning glands. I will consider just one example here. If the vesicle theory was correct, then depleting the cell of all of its secretion granules should arrest secretion. No granules, no secretion. If substantial secretion took place in the absence of granules, then there had to be another source of secreted material, presumably the cytoplasm.

We were able to almost completely deplete the cells of their zymogen granules by stimulating secretion with a pharmacological agent. Did secretion cease, or did it continue at substantial levels? With only 5 percent of the granules remaining, the rate of secretion was still highly elevated, some *10 times* unstimulated controls, and about one-third of the *maximal* response seen when the cell was filled with granules.

The vesicle model also predicted a linear relationship between the loss of granules and the amount of material secreted by the gland. If 10 percent of the granules were lost, it should produce half the secretion seen for the loss of 20 percent of the granules, and a loss of 20 percent should produce half the secretion seen for a loss of 40 percent, and so forth. This was not the case. Granules disappeared at a greater rate than their contents appeared in secreted fluid. Where did this material go? We suggested that it went into the cytoplasm.

This possibility was not widely embraced at the time, though no plausible alternative explanation was offered. If secretion continued at high rates even though only a few granules were left, they were good and swift, and though not many in number, nonetheless accounted for secretion. As for the cell-fractionation study in which tissue from active and quiescent states was compared, it was thought that the stimulant had altered the character of the homogenization and separation process in some unknown way, by chance in the right direction.

There was no evidence supporting these ad hoc explanations, nor was it necessarily clear how one would go about obtaining it. As such, we might expect the burden of proof to rest on those who argued in their behalf. Wouldn't they have to demonstrate the aptness of such arguments? But belief in the vesicle model and its underlying premises was so strong and so resolute that ad hoc hypotheses that retained its exclusivity, however unsubstantiated or unprovable, were seen as being explanatory. And if not explanatory, then at least as providing sufficient reason to reject the possibility of an alternative route. After all, they might be correct. And while it was true that to accept them would be to add further assumptions to an already loaded system of assumptions, there was no question that protein movement within and out of cells was a complicated business, and there was every reason otherwise to think that the vesicle theory was correct. In sum, there was abundant confidence that in the final analysis the ad hoc hypotheses would be substantiated and the vesicle model salvaged.

Constitutive Secretion

Let me show how this worked with another example from the period: an ad hoc hypothesis that has had a signifi-

The Tests

cant impact on the way that secretion is viewed today. It provided the vesicle theory with a kind of catchall explanation for data that did not seem to fit.

With greater rates of protein secretion, there are fewer granules in the cell. In terms of the vesicle model, there would therefore be less protein in the "pipeline." As a result, new protein would pass through the cell more quickly. We can think of a shorter queue of granules, and hence a shorter wait to exit. Or, alternatively, as fewer granules trying to leave via a set number of exit doors, reducing the crush to enter them. But however it is conceptualized, when the rate of secretion is elevated, depleting granule stores, new protein should appear in secreted fluid more quickly according to the vesicle model. This was axiomatic and it was hard to imagine anything else happening. Therefore what was observed was truly startling.

Not only didn't new radioactively labeled protein appear in secretion more rapidly, its rate of appearance was greatly *diminished.* How could this be? Somehow increasing the rate of secretion overall had retarded the secretion of new protein. There was an additional curiosity. In the absence of stimulation, the opposite was true—secretion of new protein seemed to be favored over old protein. In some unknown manner granules containing new material pushed their way to the front of the line, bypassing previously manufactured and stored protein.

The vesicle theory could not easily account for either of these observations. And after all, old vesicles had once been new. How did they get to be old if they bypassed older ones? It all seemed very confusing. But there was one important clue. Even though new protein was favored in the absence of stimulation relative to its less advantageous treatment in the presence of stimulants, this turned out to be a quantitatively trivial occurrence. New protein still accounted for only a small percentage of the total amount of protein secreted by the gland, on the order of 1 percent, even when it was *favored*. That is, most secreted protein was old, and most new protein remained in the cell to eventually become old. All that was involved was a small aliquot of new protein.

This was clarifying, but it did not get the vesicle theory out of the thicket. What conceivable purpose could such a process serve—a bypass mechanism that involved only a trivial amount of protein? And why would it be inhibited when secretion was augmented? Toward what physiological end would such a mechanism exist? It still did not make any sense. Maybe we could just ignore the phenomenon. After all, it was only a small effect, and perhaps it had nothing to tell us about how the bulk of material was released. Perhaps it was a peculiarity of no particular significance; a side track hardly worth exploring.

But science does not allow us to blinker our eyes even to such small inconsistencies. Indeed, such odd little mysteries often open the door to new understanding, and it is an important task of the scientist to explain them as completely as possible. It is not enough to say that our theory *almost* works, that we shouldn't worry about this or that trifling inaccuracy or problem, because we can rest assured that in the fullness of time it will be explained and found to be consistent with our theory.

It became clear that the seeming oddity of these observations was tied to the vesicle theory. That is, they were peculiar only if we tried to explain them by means of a vesicle transport mechanism. If instead it was allowed that the new protein passed out of the cell via its cytoplasm by membrane transport, then the observations were easily explained, indeed predicted.

The Tests

The phenomenon had all the markings of competition for exit from the cell between old and new protein. Similar properties had been described for many membrane transport processes. Different molecules or forms of molecules [for example, new (labeled) versus old (unlabeled), as in the current case] contend for the same location, usually a binding site on a protein embedded in the membrane that is responsible for or associated with the molecule's transport across the membrane. The substance that is present at a higher concentration or that has a greater affinity for the transport site will displace the other. Consequently, its transport across the membrane is favored, and that of the other substance is inhibited. This is called *competitive inhibition,* and the observation looked like a relatively straightforward example of it.

Let me explain this idea a little further. In the unstimulated state, the concentration of old digestive enzymes in the cytoplasm is relatively low, most of it being stored in the zymogen granules. When secretion is stimulated, some of this material is released into the cytoplasm, and the concentration of old protein in the cytoplasmic compartment becomes higher, at least initially. As a result, after stimulation, recently manufactured (labeled) protein being released into the cytoplasm from ribosomes or other proximal compartments in the secretory pathway competes less favorably for release from the cell than it does in the unstimulated state, when the concentration of old protein in the cytoplasm is lower. But this difference is only relative, and even in the unstimulated circumstance the amount of old protein in the cytoplasm greatly exceeds that of new protein. Therefore, the older material predominates in secreted fluid in both cases.

That this explanation is correct was demonstrated subsequently in a variety of experiments that I do not have the space to go into here. It also became clear that these observations were not an idiosyncrasy of the particular system we were studying. They appeared to be properties of secretion systems in general. And yet, however general the property or valid the explanation, it did not gain favor. Even though we had described a property of the natural system—and it was hard to label it an artifact for that reason—the explanation we proposed just had to be wrong. What we had observed had to fit the vesicle model in some manner, somehow, because there was no alternative. But how?

Many years later, a solution was proposed. It was suggested that the small (40-nm-diameter) shuttle vesicles that were thought to carry new protein from place to place *within* the cell also released their contents directly into extracellular spaces after fusing with the cell's enclosing membrane. Such a mechanism could explain the results just described if we are not too picky about the details. One such detail is that because these objects are far smaller than zymogen granules, it would take 100,000 or more fusion events to release the equivalent of the contents of a single secretion granule, potentially making the process 100,000 times less efficient. Nonetheless, whatever its likelihood, the conclusion was that there was a vesicle bypass mechanism for new protein, not a membrane transport process, as we had suggested. This came to be known as the *constitutive pathway* because secretion via this route was thought to occur continuously, independent of the presence of external stimuli.

Thus the vesicle theory had been saved. All that was needed was another type of vesicle, one that carried new protein, had a low transport capacity, and could bypass the traditional pathway. The small vesicles fit the bill well enough, and constitutive secretion became an established aspect of the secretion mechanism. And yet, there was not then, and to the best of my knowledge there still is not, any serious evidence that demonstrates that these small vesicles fuse with the cell membrane and release their contents into the extracellular environment, much less at a sufficient rate to account for what is observed. And remember the numerous other uncertainties about these objects. Though these vesicles were widely believed to contain proteins that were to be secreted and to fuse with various intracellular membranes along the vesicle pathway, these were little more than conjectures. Most important, given that it all occurred just as the vesicle model would have it, there was no evidence that any of this actually accounted for the particular property of the intact system that I just described. What the hypothesis did was offer an explanation within the boundaries of the vesicle model, and that seemed quite enough; the small vesicles did exactly what was required of them.

The Zymogen Granule

However such evidence was interpreted, it was important to directly measure protein transport across the relevant membranes. In the final analysis, this would tell the story. Were the involved membranes permeable or impermeable to digestive enzymes? If the membrane encircling the zymogen granule was permeable, then these molecules would enter the cytoplasm and there would necessarily be a cytoplasmic pool. And if the cell membrane was permeable, then they would be secreted across it.

We studied the zymogen granule first and our initial studies were as simple as could be. The granules were extracted from the pancreas after tissue homogenization and suspended in fluid to assess whether digestive enzymes leaked from them into the medium. If leakage occurred, it would be prima facie proof of membrane permeability. If not, it would be conclusive proof that such a permeability did not exist, and that only vesicle mechanisms were possible. When the measurements were made, protein leaked out of the granules. Biological membranes, it seemed, were permeable to protein molecules—at least, this membrane to these molecules!

Even though there was certainly no way that this observation could be seen as demonstrating that the granule membrane was *impermeable* to its protein contents, the simple fact of release was not in itself sufficient to affirm the presence of a natural permeability. Artifactual explanations were possible. Remember that in order to obtain the granules, we had to homogenize the tissue. This meant that the observations were susceptible to homogenization-induced artifacts of the sort we have already discussed. Even though the granules looked intact in electron microscopic images, we couldn't be sure that the procedure had not ruptured some membranes and that it was the contents of these damaged granules that had access to the suspending medium. Or perhaps digestive enzymes from other cell components had been bound or adsorbed to the surface of the granule membrane during homogenization, and it was this material that was being released into the medium. Or perhaps both occurred.

To determine that the membrane was permeable, we had to exclude these potential artifactual explanations. Although we could not know how many granules might have ruptured or how much protein had been adsorbed to the surface of the membrane, we could remove these potential sources of artifact simply by washing the granule sediment in a large volume of fluid *prior* to carrying out the experiment. When this was done, the washed granules also released their protein contents.

There is little doubt that this would have been evidence enough to establish the permeability of a biological membrane to most substances. But there was such a strong predisposition against a permeability in this case that, although the observations were direct and convincing on the face of it, artifactual explanations continued to be sought. The possibility remained that washing had somehow damaged additional granules, and that we were still dealing with an artifact of separation. Indeed, we could wash and wash the granules to our heart's content, treat them in this protective way or that, but the possibility would always remain that some new artifact had been introduced by whatever operation we performed. Maybe the more we fooled around with the granules, the more artifacts would be produced. There was no evidence that this was the case, but the possibility was not easily excluded. To convincingly demonstrate to a disbelieving audience that the membrane was permeable to protein, we had to show more than the simple occurrence of release. We had to understand its properties.

As already discussed, a central characteristic of artifacts of homogenization and separation is that they are of a fixed quantity. The specific procedures used might damage a certain percentage of the granules, or lead to a certain amount of protein adsorption. The sum of these artifacts would be the fixed amount of protein that could be released into the medium. If they accounted for 1 percent of the protein in the suspension, then 1 percent could be released; 5 percent, 5 percent released; and so forth. But whatever their value, if such artifacts were the sole cause of release, the remainder—the contents of undamaged granules—should remain in place, unperturbed by resuspension.

This understanding led to four predictions. First, self-evidently, as long as the method of preparation remained unchanged, the total amount of protein *capable* of being released into the medium would be constant from sample to sample, within measurement and statistical error. Second, if the method of preparation was satisfactory, this value would not be the whole contents of the granule suspension (100 percent), but some defined, probably relatively small, fraction of it. Third, if we removed all of the material that could be released in these artifactual ways, no further release could take place. There would be no other potential source. And finally, since release as a consequence of rupture or adsorption is not constrained by a (membrane) barrier meant to restrict protein movement, we would expect it to occur almost instantaneously on resuspension of the granules, within the few seconds required to create the suspension by mixing the granules with a large volume of fluid.

The predictions differ if the membrane is permeable to the proteins. In this case, the amount capable of being released is not a fixed value, but a variable proportion of the whole. It should be possible to go from zero release to essentially 100 percent, simply by our choice of circumstances. We should be able to limit release to small amounts, completely deplete the granules of their contents, or set it at any level in between by our manipulation of the system.

Second, if we removed protein that had already been released into the medium, the remainder within intact granules would still be *available* for release, not inaccessible behind an impermeable membrane. The unloading of granules would be expected to continue despite the removal of already discharged protein unless and until the granules were fully depleted of their contents. And finally, simply mixing the granules with the suspending medium would in itself have little effect on release initially. It would take time for a substantial quantity of protein to cross the restrictive membrane barrier. Although how slowly release would occur could not be predicted a priori, it would be far slower than in the absence of a membrane barrier.

These predictions were all based on the laws of diffusional equilibria that govern the movement of molecules across short distances. In such equilibrium systems, the substance moves (diffuses) in both directions across a plane or barrier. For complex chemical-transport systems like those found in biological membranes this does not mean that movement necessarily occurs with equal facility in both directions. Nonetheless, if membrane transport takes place, we would expect some portion of the released protein to reenter the granules if given the opportunity. Because movement would be driven by concentration, as more is released and the concentration of protein in the medium increases relative to that inside the granule, the amount reentering would also increase in direct proportion. Eventually, as the concentration in the medium rises, the rate of entry would come to equal the rate of exit. When this occurs, we say that we have reached the *equilibrium state*. If one side continues to contain the substance at a higher concentration than the other, even though movement in and out of the object are occurring at the same rate, we say that we have reached a *steady state*. In either case, for every molecule that leaves, one enters, and there is no longer any *net* movement of material into the medium from the granules as seen at the outset. When this happens, the concentration of protein in the medium and the protein content of the granule become constant; that is, they no longer change over time.

With these understandings in mind, we tried to distinguish between the predictions of an artifactual explanation for release and a membrane transport explanation. For example, samples of the same size (the same number of washed granules) were resuspended in widely differing volumes of medium. If the membrane was impermeable to protein, then the different volumes of suspending fluid should have had no noticeable effect on release as long as they were large enough to displace all remaining artifactual material from the particulate phase into the medium. If, on the other hand, the membrane was permeable, increasing the suspension volume would increase the amount of material released proportionately, because the larger volume could accommodate a greater amount of secreted protein at a given concentration. At large enough volumes, the granules would be wholly depleted of their contents. We found that this was the case. We could vary the amount of protein released from the granules simply by our choice of volume. This was exactly what was expected if release occurred as the result of transport across a permeable enclosing membrane.

We also found that when the released protein was removed from the medium (by filtration), the intact granules did not stoically hold on to their remaining contents, but released more. Not only that, but one could vary the rate of release by varying the rate of removal. If the protein was removed quickly enough, a maximum rate of release was reached that could be maintained over extended periods of time. All of this was as it should have been if the membrane was permeable to the proteins inside the granule. If it was not permeable, release would have fallen off exponentially as the artifactual contents were washed away.

This brings us to the final prediction: the rate at which release occurs. We found that the release of proteins could be set to occur quite slowly, with half-times of the order of an hour. Not only that, but in this circumstance release was greatest at early times, then slowed down, and eventually ceased as the concentration of protein in the medium reached its maximum value. All this was in accordance with a membrane transport model. If rupture and the release of adsorbed material had been the sole causes of discharge, protein release would have been complete within the time required to mix the granules with the medium—a matter of seconds.

Granule Permeability as Seen in the Cell

Taken together, these observations provided strong evidence that the granule membrane was permeable to digestive enzymes. In each and every instance, the properties of release were those predicted for a membrane transport process, and if artifacts of preparation were present, their effect seemed minor. Still and all, the granules had been examined after their isolation from the cell. Could we be sure that the same features would be observed in the intact cell? Perhaps the method of separation had made impenetrable membranes penetrable in some fashion. Though there was no reason to expect this, it was nonetheless important to demonstrate the same or analogous properties in the whole cell, as I have been admonishing.

We attempted this by marking the path in the other direction—by following the enzymes as they moved backward from the medium suspending pancreatic cells into their cytoplasm, and then into the granules. This would not only confirm that cell and granule membranes were permeable to digestive enzymes, but would illustrate the reversibility of the transport process. The vesicle theory proposed that transport both within and out of the cell took place as the consequence of an irreversible sequence of events that is usually called *vectorial transport.* In vectorial transport, material moves from the site of synthesis through the various compartments in the vesicle pathway, solely in the forward or secretory direction.

We used a radiolabeled digestive enzyme to trace the process. This had an important advantage. It was possible that uptake would occur as the result of an irreversible vesicle mechanism of some sort, not membrane transport. This is called *endocytosis*, to contrast it to exocytosis. Endocytosis is like exocytosis in reverse. The substance would enter the cell in small vesicles formed from the cell membrane. First it would be captured and then encapsulated as the vesicle is folded into the cell.

If this were the case, the labeled protein would be constrained to the small endocytic vesicles, and only small amounts of material could be accumulated in comparison to the large stores of unlabeled endogenous digestive enzyme in the cell. Accordingly, the radioactive protein would be greatly diluted by the cell's unlabeled digestive enzyme stores, reducing the *specific radioactivity*—the ratio of labeled to unlabeled material—as compared to that in the incubation medium.

On the other hand, if uptake took place by a membrane transport mechanism, then eventually the specific radioactivity of the enzyme within the cell would approach that in the medium. That is to say, rather than being diluted by endogenous pools, it would equilibrate with them. Exogenous (labeled) and endogenous (unlabeled) protein would move reversibly across the relevant membranes, exchanging with each other. Over time an equilibrium between the two would be achieved, and the material in the cell and the medium would have essentially the same specific radioactivity—the same ratio of labeled to unlabeled protein.

This was all well and good in theory, but to actually do the experiment was not a simple matter at the time. Because we were interested in the location of the labeled protein in the cell, we turned to microreductionist manipulations. We chose to homogenize the tissue and separate its various fractions. Defined subcellular fractions, the most important being the zymogen granule and cytoplasmic fractions, were collected and assessed for the appearance of labeled protein. The first question, of course, was whether the labeled protein entered the cell at all. If it did, was it greatly diluted or did it equilibrate? Was all of the labeled material recovered in small (endocytic) vesicles, or was it in the cytoplasmic fraction and the zymogen granules? And if the latter was the case, did it appear in the cytoplasmic fraction before entering the zymogen granules?

To make a long story short, things were just as the membrane transport model predicted. Uptake by tissue was observed, and the labeled material was not restricted to a sediment of small vesicles, but was found mainly in the zymogen granule and cytoplasmic fractions. First it appeared in the cytoplasmic fraction and then in the zymogen granules. And finally, and most important, after a while the specific radioactivity of the suspending medium, the cytoplasmic fraction, and the granules came to approach each other, though such equilibration was not seen for other fractions of the cell, such as the mitochondria or microsomes. This provided strong confirmation that both cell and granule membranes were permeable to these proteins, and that transport occurred into and out of a cytoplasmic pool.

Nothing but Artifacts

Nonetheless, when we performed these experiments in the early 1970s, both with isolated granules and whole cells, what we observed was understood to be impossible. The idea that these or any other proteins moved through membranes in an analogous fashion to small organic molecules and ions was unacceptable to most knowledgeable scientists. And so it was not surprising that subsequent to our first report of the phenomenon in *Nature* in 1972, and for many years thereafter, it was argued that whatever we had seen, could only have been the consequence of some sort of artifact.

The rupture of membranes was the most common explanation for the in vitro observations on zymogen granules. This was referred to as *granule lysis.* It was believed that granule breakage caused release in one way or another, whatever the experiment, whatever the result. If release was slow, then granules broke open infrequently. If it occurred rapidly, lysis occurred rapidly. If it was increased by enlarging the suspension volume, then the larger volume had somehow increased lysis. And if we could maintain release at high rates simply by removing the released protein from the medium by filtration, then this act must have somehow increased granule lysis.

Though the evidence argued against it, the ad hoc hypothesis that release in each and every instance was the consequence of granule lysis appeared to have far greater valence than our experimental data. The burden of proof was widely perceived as resting squarely on us; it was our obligation to prove that rupture did *not* account for the results. We would have to prove the negative.

Seeing Is Believing

Whether such a conclusion was appropriate or not, fair or unfair, scientific or unscientific, the simple fact was that if one hoped to convince the unconvinced, it had to be affirmatively demonstrated that granule lysis did not account for protein release. But the experiments had done just that. They demonstrated that granule lysis did not account for the results with isolated granules, and certainly not those obtained in situ.

The Tests

The conclusion that the membranes were permeable was not only a reasonable interpretation of the data, not only the most direct and convincing explanation, but the only one supported by the evidence. But this changed little. If the skeptics were to be convinced, something beyond the kinetics of transport was needed. Something more concrete, more obvious—even, if possible, more direct. If one could only see an intact granule actually releasing its protein contents. As they say, seeing is believing.

This was not feasible at the time, but some 20 years later technical advances had made it possible to definitively assess what happens during protein release from these tiny objects. Individual zymogen granules could be observed with a high-resolution microscope while they lost their protein contents. The first measurements were made in the late 1980s, using an elaborate x-ray microscope at the Brookhaven National Laboratory on Long Island. It had become possible to produce very bright coherent beams of soft x-rays (x-rays in the 20- to 40-nm wavelength range) from synchrotron sources and focus them to a very small spot. This produced an x-ray microscope with sufficient resolution to clearly image zymogen granules at high magnification. Unlike the electron microscope, the x-ray microscope allowed the sample to be examined whole, in the absence of a vacuum and while suspended in water. Furthermore, because the wavelength of the incident x-rays could be tuned, it was possible to choose light of wavelengths that were preferentially absorbed by protein, thereby furnishing a quantitative measure of the protein contents of the granules. Taken together, these features—high resolution, imaging whole objects in aqueous media, and the ability to measure the protein contents of individual granules—made it possible to observe these objects as they were releasing their contents and to make quantitative measurements of the process.

Rupture did not occur. Granules were not seen full at one time, only to disappear after having broken open and disgorged their contents. Instead, they *remained intact* while releasing their protein contents. Release occurred gradually from apparently intact granules and simultaneously from granules suspended in the same environment. As the flow of medium through a specially designed chamber removed previously released material, the granules released more, just as in the filtration experiment. And we could even fill the granules, simply by placing high concentrations of digestive enzyme in the suspending medium.

The Irony of It All

In the 1960s, at about the same time that we discovered non-parallel secretion and began exploring the possibility of a membrane permeability to protein, a major revolution was under way in how biologists thought about the membranes of cells. The lipid bilayer model I described earlier was in disarray. It had been thought that protein molecules could only be associated with the membrane's surface. If they penetrated the bilayer, they would necessarily be in intimate physical contact with nonpolar aspects of the membrane's lipid constituents, and the traditional belief was that this was not possible because of their polar nature. But now evidence indicated that proteins could be buried deep within the membrane's nonpolar core. Indeed, some seemed to completely transect the membrane, forming aqueous channels and complex transduction structures to carry matter and information in and out of the cell.

Moreover, it turned out that the bilayer had qualities of both a liquid and a crystal. Like a crystal, it was an ordered structure. The lipid molecules were lined up like soldiers standing back to back in two oppositely oriented rows, the bilayer. But its molecular constituents were not fixed as in a cellophane membrane. They moved with great rapidity to and fro both within and across the bilayer's extremely thin substance, along with many of its proteins, just like liquid molecules.

Nor could the bilayer be modeled as a bulk nonpolar phase in which proteins had to dissolve before they could pass. It was only two molecules thick, more a surface than bulk material. And significantly, many of the lipid molecules had two essences. They were not simply nonpolar, but had polar aspects as well. Such molecules are called *amphiphilic*—substances of two desires. Many membrane lipids were amphiphilic, with polar head groups that faced the aqueous environment, and antipodally situated nonpolar parts, such as long hydrocarbon tails, that faced each other away from the water. Although this had been known for a long time, its significance in regard to the possibility of protein transport had not been appreciated. All proteins had to do was to cross this very thin, very mobile, very easily influenced barrier of ambivalent physical character. This was not necessarily an easy task, but it was not the impossible one that had been imagined.

It also turned out that proteins were not as emphatically polar as had been thought. Although usually quite soluble in water and insoluble in nonpolar media, proteins are commonly formed from roughly equal numbers of polar and nonpolar amino acid subunits. Like the membrane lipids, they were amphiphilic molecules. When placed in aqueous media, they would show their polar face to ease their association with water and would become soluble as a result. At the same time, they would hide their nonpolar elements as best they could in protected pockets within the molecule's interior, away from the water. As a consequence, in aqueous solution most proteins present a roughly spherical surface dotted with polar groups. Any protein—indeed, any molecule—in this situation would do the same. It would attempt to maximize its potential for interactions with water, putting its best, polar face forward, and hiding its incompatible, less attractive nonpolar elements. It was this proclivity that made proteins polar substances.

But molecules do not just acquiesce to water's need, forming pure polar balls at all costs. Their structure determines what accommodation is possible. For proteins, this accommodation is limited by the particularities of the covalently linked sequences of amino acids that form each peptide chain. Depending on where in a chain a particular amino acid is located, it may be fated to the surface of the molecule or its interior regardless of its polar or nonpolar attributes. As part of a chain, an amino acid may be forced to the surface or relegated to the interior of the molecule by the properties of its attached neighbors. The only way to fully accommodate to water's claim would be for the chain to break apart, to self-destruct. Short of that, the molecule as a whole conforms, however uncomfortably, to the configuration that is most stable in an aqueous environment. This results in a structure that is at its lowest free-energy state for that particular environment, and this usually leaves numerous nonpolar amino acids on the molecule's surface and polar residues facing inward. This is the best it can do.

The same applies to molecules in a nonpolar medium. They will attempt to accommodate to this environment as well, changing shape and the location of polar and nonpolar aspects as best they can within the limits imposed by their unique structural constraints. However easily or with whatever difficulty, all molecules attempt to adapt to their environment in this way. And yet in spite of this understanding, for most of this century protein molecules were seen as wholly inflexible, rigid polar balls of essentially singular and invariant structure, unable to make any accommodation to a nonpolar milieu.

With the hundreds of amino acids that form the peptide chain having to interact with each other in three dimensions to form the finished folded and functional protein molecule, their structure seemed like a puzzle that could have only one workable solution. We know today that this characterization was wrong. These most complex of molecules are capable of significant, and sometimes remarkable, structural adaptability in the face of changes in their environment. To characterize proteins as polar molecules, in the same way that we consider simple electrolytes and small organic molecules polar molecules, was naïve. In a very real sense, proteins composed of the same amino acid sequence in different environments can be viewed as totally different molecules, if by *different* we mean that they display distinct physical and chemical properties.

Passage

If proteins are found within the membrane bilayer, then obviously they must have entered it in one way or another. And if they can enter it, then they must be able to exit it as well, at least if the laws of chemical and diffusional equilibria apply. That is, in the ordinary course of things, such processes would be expected to be reversible. Assuming this to be the case, couldn't the molecule exit across the other side of the membrane just as well as across the side it entered? And if exit could occur across either surface, then a protein entering the lipid bilayer could be transported across it.

This conclusion about the possibility of membrane transport for protein molecules was reinforced by another discovery. Pores large enough to accommodate even the largest proteins were found in the membrane of the nucleus. In some cells these membranes seemed to be littered with large holes. Thus, two discoveries—proteins are embedded in the bilayer and pores large enough to accommodate them exist in at least some membranes—called the two assumptions that had excluded membrane permeability to proteins for so long into serious doubt. Indeed, the assumptions seemed completely misconceived. One could not presume a priori, as had been done, that proteins could not pass through the lipid layer because the bilayer is nonpolar and proteins are polar. Some water-soluble—that is, polar—proteins might be able to do just that. Nor could one assume that aqueous channels in membranes were too small to accommodate protein molecules. Ones large enough were known to exist.

This meant that some membranes might be permeable to particular proteins. But there was also the prospect that permeability to protein molecules might be a rather general feature of cells and biological membranes. At the least the existence of such processes could no longer be excluded on the basis of a priori assumptions. The question had to be asked protein by protein, membrane by membrane. At the time these discoveries about membranes were made, there was precious little evidence, other than our own work, on the subject of direct protein transport across membranes.

In light of these findings, one might have expected our evidence to gather at least a patina of credibility. Membrane permeability to protein seemed to be a legitimate hypothesis. And why not in the exocrine pancreas? But this turn of mind did not occur, other than in the sentiments of a few isolated individuals. It became the common understanding that though certain proteins, membrane or membrane-associated proteins, could exist in intimate contact with lipids in the bilayer, and as such had obviously entered the membrane, they could not cross it.

It was believed that the only way that proteins in membranes could be removed from them by the cell was to remove and degrade whole portions of membrane en masse as the result of the formation of small membrane vesicles. Simple physical equilibrium between the membrane's protein contents and the surrounding medium did not occur, and hence their transport across its substance was not possible. As for the large pores in the nuclear membrane, they were understood to be a singular exception, unique to this structure where they served its very special purposes. Most nuclear proteins were manufactured in the cytoplasm of the cell and had to enter the nucleus subsequently, and the large messenger RNA molecules that were produced in the nucleus had to leave it to participate in protein synthesis in the cytoplasm. This was why there were large pores in the nuclear membrane, and while it was true that various proteins and nucleic acids moved in and out of the nucleus through them, membranes porous to large molecules such as proteins did not exist otherwise.

A Most Important Paradox

But there seemed to be a paradox. As we have discussed, it is on ribosomes attached to the membranes of the endoplasmic reticulum that the synthesis of proteins that are secreted by the cell—its exportable proteins—is thought to take place. Curiously, these ribosomes are affixed to the external, not the internal, surface of these membranes. This seemed odd, because for newly manufactured protein to travel through the cell by the vesicle pathway, they must be *inside* the endoplasmic reticulum, in its cisternae, not outside them. As we have already explained, to get into the internal spaces of the endoplasmic reticulum the protein first has to cross its membrane. That is, the vesicle model required that proteins transported in vesicles *because they cannot directly cross membranes* first *have to* pass through a membrane. What an extraordinary contradiction! The vesicle transport's raison d'être was that proteins could not cross biological membranes. And yet for vesicle transport to take place, they seemed to have to do just that.

Not long after this realization, an hypothesis was devised to explain how this occurred. It is called the *signal hypothesis,* and soon after its proposal a variety of evidence was published indicating the presence of such a mechanism. Just recently a Nobel prize was awarded to Gunther Blobel for its invention. This was the first explicit proposal of a particular mechanism for the membrane transport of proteins and it was consequential. It came to be appreciated that membrane transport processes for proteins existed. The unimaginable, ignored, and avoided had at last found a mechanistic incarnation. Soon it came to be understood that numerous proteins, in all kinds of cells, crossed a wide variety of membranes by membrane transport, not vesicle transport. For example, bacteria secrete proteins, toxins, enzymes, and so forth, and yet they contain no vesicles. Many enzymes in mitochondria, crucial to energy metabolism, are synthesized in the cytoplasm and have to enter the mitochondrion across its enclosing membrane, and there was no indication that vesicle mechanisms were responsible for this. And of course there was the nucleus. In any event, over time the list of membrane transport processes for proteins became substantial.

But the mechanism the signal hypothesis proposed was remarkably restrictive. It allowed a given protein to cross only one particular membrane, only once, and only in one direction (for example, from outside the reticulum to its cisternae). The reason for this limitation was that membrane transport was envisioned as occurring exclusively during an unique event—the synthesis of the peptide chain. It happened as the peptide chain was being formed on a ribosome attached to the membrane. As already noted, according to the signal hypothesis, the ribbonlike peptide chain snaked its way through a small membrane pore that sits directly under the ribosome. Having passed through the pore, the new protein folds into a tight polar ball, unable to escape the cisterna, or for that matter cross any biological membrane ever again.

The signal hypothesis was an exceptionally complex model for molecular transport, and also an extremely confining one. It had chiseled a chink in the armor of the impermeability of membranes to proteins, but only a very small one. It restricted membrane transport for proteins in the most severe manner possible, though still allowing for its occurrence. It was as if one had at one and the same time proposed a mechanism for protein transport and denied its existence.

It was hoped that this mechanism would provide a universal solution for the problem of membrane transport for proteins wherever it occurred, whatever the protein, whatever the membrane. In the end it failed to do so. Even though there was a great deal of evidence consistent with the signal hypothesis, from the earliest studies other means of transport were also indicated. Today, the evidence suggests that a mechanism something like the one proposed in the signal hypothesis may occur across the membrane of the endoplasmic reticulum, although many of its details remain unclear and the important quantitative proof that we discussed earlier is still lacking. But it is not clear that such a mechanism is found anywhere else.

More important, the tight corset that the signal hypothesis imposed on the membrane transport of proteins was unable to contain the abundance and variety of these processes. Today we know that there are many mechanisms for the transport of proteins across membranes, most of which occur *after* protein synthesis, not during it. Sometimes they pass through the bilayer, sometimes through large pores in membranes, and sometimes they pass by means of special mechanisms that are similar in kind, although different in their specifics, to chemically mediated membrane transport mechanisms for small molecules.

Over the past decade or so—some 30 years after our original observation of membrane transport for protein molecules, and also some 30 years after the old bilayer model for the structure of the biological membrane was revised—it has become clear that membrane transport occurs for all kinds of proteins, in all kinds of cells, across all kinds of cellular and intracellular membranes, by a great variety of mechanisms. Not only do such processes exist, but, as I have said, they are central to life. They are in the main what accounts for an event that is essential to the living state—the topological partitioning of protein molecules in cells.

In Spite of Everything

Yet in spite of everything I have written here about the vesicle theory and membrane protein transport, I believe that it would be more than fair to say that most biologists, and certainly most experts in this area, still believe that the mem-

The Tests

branes of the acinar cell are absolutely impermeable to the digestive enzymes. It is still held that the mechanisms proposed in the vesicle theory have been established as their sole modality of transport, and for that matter for the transport of secreted proteins in virtually all plant and animal cells. Even with our modern understanding of the biological membrane and protein structure, the widely appreciated occurrence and importance of membrane transport for protein molecules otherwise, the long-standing evidentiary weaknesses of the vesicle model—the evidence I have discussed for direct transport, and much more that I have not had the space to discuss, little has changed over a period of some 30 years.

For those open to the prospect, there was clear and convincing evidence 30 years ago that certain membranes of the pancreatic acinar cell were permeable to digestive enzymes. And it was equally clear that the assumptions that gave the claims of the vesicle theory their gravitas could be put to the test experimentally, but for the most part were not. Nor, given the existence of a vesicle mechanism, was there any quantitative proof regarding the extent and nature of its role, either in and of itself or in relation to alternative membrane transport mechanisms. Nor was there any reason to think that both types of mechanism could not exist in the same cell, side by side, and even move the same substances.

Why, in spite of all this, was there and does there continue to be such a deep and unswerving belief, not only that a vesicle transport mechanism like the one I outlined exists for protein secretion in animal and plant cells, but that it fully accounts for this process wherever it occurs? Nothing, it seems, neither evidence nor ideas, has been able to weaken this strongly and earnestly held belief. It has remained the dominant paradigm.

CHAPTER 13

The Call to Authority

> *Frantic orthodoxy is never rooted in faith but in doubt.*
>
> Reinhold Niebuhr

Why has there been so little doubt about the vesicle theory over the years, when there should have been so much? I have tried to demonstrate why, by examining the prominent role of the strong-microreductionist principle—the principle that the parts entail the whole—in that belief. I hope that you can now see that it was the strong-microreductionist approach to learning about nature that gave rise early on to unwarranted confidence in the vesicle theory's dominion. And how it came to be that research sought to explicate, but not challenge, the model's soundness. Or why critical tests of the theory against the properties of the intact system seemed unnecessary, if not totally pointless. Indeed, why the natural process was hardly in evidence, and knowledge of it thought superfluous. Nor, finally, should it be hard to comprehend why the evidentiary weaknesses of the vesicle theory were not, and are still not, adequately appreciated.

Research in this area has been much like building a magnificent castle in the sky. While a glorious achievement of the imagination, it has unavoidably lacked a foundation in the rich, dark soil of the earth. Although imagination is the most magnificent instrument of human cognition, indeed of life itself (certainly no microreductionist construction that), we should not confuse how we imagine things—our theories, hypotheses, and models—however luminous, with knowledge of nature.

Sanction

In a review in the journal *Nature* of a technical book that I wrote on this subject about 15 years ago, the author made two major points. First, he accused some of the inventors of the vesicle theory, as well as its promoters and camp followers, of promoting a cult of personality. Such a cult, he argued, leads to authoritarian and dogmatic posturing. Having excoriated them, he then took me on. He said that while my criticisms of the vesicle theory were in the main justified, the evidence I presented as proof for membrane transport did not convince him. Accordingly, or so he seemed to conclude, the status quo ante obtained. Since I had not convinced him otherwise, whatever weaknesses there were in the evidence for the vesicle theory were immaterial. One could still have confidence that it described the way things actually happened. The reviewer did not give reasons for his demurral. Nor did he tell his readers exactly what he found unconvincing in the experiments I had discussed, or on what basis he was able to dismiss them. Of course, he was entitled to his opinion, but for criticism to have meaning beyond personal opinion, it must be justified by precise and reasoned argument. Certainly, if this is true anywhere, it is in science. Nonetheless, the reviewer thought his opinion, his mere opinion, sufficient.

The Call to Authority

What he had done was just what he had accused proponents of the vesicle theory of doing. The authority of his own opinion seemed enough in itself to dismiss whatever evidence had been put forward to the contrary. While complaining about authoritarian attitudes, he had given them expression. Such calls to authority have permeated this type of biology from its origins in the traditional histology of the nineteenth century. The study of the microscopic appearance of cells and tissues was as much traditional naturalist observation (writ small, that is, microscopically) as it was science. But naturalist reportage without interpretation is useless to the scholar who seeks to understand. As a consequence, the expert observer who described a phenomenon also offered an interpretation. And such opinions, however hypothetical, however justified or unjustified, came to bear a great and often determinative weight.

Today the tools of the trade in cell biology are among the most advanced that modern technology has to offer. But tools are not enough, and never will be. And the old ways die slowly. The way in which we gather and analyze data is crucial to the quality of our understanding. And I am not just referring to careful measurements, good controls, and proper statistical analysis, as important as they are. Our experiments, the results we obtain from them, and the meanings we attribute to those results are only as good as our approach to learning about nature, however carefully performed the experiment may be otherwise. In the end, our *philosophy of learning* is crucial, by which I mean how we organize, examine, and evaluate our *ideas* and their cogency. After all is said and done, this so-called scientific method is not some bothersome hindrance to acquiring knowledge or to performing important, practical research, that we best ignore; nor is obeisance to it demonstrated by a scientific Hippocratic oath taken by each recipient of a degree in the sciences. However inconvenient it may seem to the working scientist, and however adequately or inadequately it may be applied, it remains essential. It cannot be ignored, because our efforts are ultimately measured against its demands.

As so it must be for the study of secretion and the transport of large molecules such as proteins. Because the strong-microreductionist mind-set has dominated thinking about these subjects for the past 50 years, critical analysis has been notable by its absence. As discussed, the vesicle model was not tested against the natural process it hoped to explain, because the strong-microreductionist attitude engendered confidence that a full understanding of the mechanism could be captured by inductive inferences from knowledge of what appeared to be its component parts. Such a process can never be justified, but its application to the study of molecular transport was particularly egregious. The bedrock in the study of the movement of molecules, for good and substantial reason, is the actual quantitative measurement of that movement as it is found in nature. Without such knowledge there is really nothing to study.

By deciding that it was unnecessary to describe as fully as possible the actual transport process for proteins in order to uncover the underlying mechanisms, the practicing scientist was ultimately and inescapably thrown back on accommodation or acquiescence to authority of some sort. That authority might be personal—the scientist's confidence in his or her own beliefs—that of an esteemed expert, or the additive power of a community of believers; but it necessarily reflected acquiescence to authority of one sort or another, and not to the facts of nature. Such beliefs are by their very nature dogmatic. And while dogmatic belief is an occupational hazard for scientists, for those with a strong-microreductionist bent it is unavoidable, because they have determined that a particular human construction of the natural world can serve as the arbiter of truth in lieu of nature itself.

It is self-evident that scientists have a deep need to understand. This seems to be why many chose their profession. But along with this need there often comes the compulsion to view one's particular sense of understanding as being clear and certain, just as with many devoutly religious people. After all, is it not the scientist's task to rid us of ignorance and uncertainty, and replace it with clear and unequivocal knowledge? But doubt is as much a part of science as is knowledge. And ironically, the search for understanding is always inevitably and doggedly shadowed by ghosts of doubt, uncertainty, and ambiguity. We cannot rid ourselves of them. For when they disappear, so does science.

But strong microreductionists have no way to accommodate such contingent beliefs. Instead, they gain confidence that they are providing a true and accurate description of nature as they adduce evidence consistent with their theory. As we have seen, such evidence is often easy to come by and can give the illusion of comprehension, however unearned. In this way models can become totemic, like a child's security blanket, and not merely devices of practice. And, as with children and their security blankets, belief in their protective power may be hard to put aside.

When the widely accepted beliefs outlined in the vesicle theory were confronted with the actual properties of the intact transport system, it was those properties, and not the theory, that came into question. It seemed virtually impossible to obtain results that could not be interpreted in some way that was benign to the vesicle theory by those who had a mind to do so. Inconvenient observations could always be explained by ad hoc hypotheses of one sort or another, however lacking in evidentiary support they might be. And one could always raise the ad hominem criticism, that whatever the evidence, the bothersome work had not been done carefully or completely enough to be accepted as such. Even when this strategy failed, it was usually sufficient simply to hold on to the vesicle model by arguing that more and better contradictory evidence was needed before a change of heart and mind was warranted.

Seeking True Understanding

As already noted, Karl Popper said that in spite of the crucial role of falsification, one should not reject a good theory just because of a few negative results. True enough, but if we are not willing to reject a theory as a consequence of falsifying tests, then on what basis can we reject it? That is to say, if we consider such tests contingent, then how can the truth be apprehended? Since it would be extreme to insist on rejection as the result of *any* falsifying test, we are left with the status quo. Though research presents us with facts, and our powers of reasoning provide us access to logic, what is believed in many areas of what we call science—biology providing many good illustrations—rests ineluctably on the ruins of the old order, where authority and community opinion serve as the guides. And strong microreductionism can contribute, as in the current case, a sturdy philosophical mortar to bind the bricks of authority and theory into a coherent but fundamentally flawed system of understanding.

The hope of science in this circumstance is that in time the truth will win out. This is called *Planck's law*. Max Planck, the great physicist, indelicately said that scientists do not change their views—they simply die, and new scientists without the same ingrained prejudices take their place. This

The Call to Authority

proposition is simultaneously pessimistic and optimistic. It is pessimistic because it offers no hope that people will change their minds. It is optimistic because it assumes that the new generation will not be terminally "infected" by the beliefs of their teachers and elders.

The reality, however much we would like to deny it, is that how things get resolved in science is a complex and uncertain business. Unquestionably, the scientific method and the character of the evidence play a crucial role. They are hardly irrelevant, as some critics of science argue. But these critics are correct in pointing out that scientific belief is not merely a matter of facts and reason. Prejudice about ideas and people; personality; the power of authority and prior belief; raw political power; who controls journals, organizations, and funds; the depth of commitment to an idea; and any and every other human and social attribute and foible that one can imagine are also at play. And the outcome, though not the truth itself, is a community decision made either by affirmation or acquiescence. Either the community of scholars buys into a notion, or it does not, for both good and bad reasons.

It is the task of the scientist to explain nature as it exists, not as we will it. And this can be done only by confronting our theories and notions with as clear and unambiguous observations of nature itself as we are capable of making, and by being willing to put aside, even if painfully, our most cherished notions if the evidence requires that of us. Strong microreductionists excuse themselves from this central task of science. It is for this reason above all else that the conviction that the parts entail the whole is inadequate, not only to describe life, but as a strategy for learning about it.

The faith that we will fully comprehend the actual workings of life's genuine machinery if we are able to elucidate the inmost parts of what we believe to be its mechanisms is as incoherent as it is commonly applied. Yet curiously, the strong-microreductionist belief, though resolutely held, is rarely expressed as such. There is good reason for this. Affirmatively, it has seemed to be spectacularly successful. This has not only been its best advertisement; it has left scientists with little motive or inclination to note, much less mistrust, its implications, effects, or authenticity. As I have noted, whether acknowledged or unrealized, forgotten or ignored, the seeming success of the strong-microreductionist approach represents a mistaken judgment. Such successes can invariably be shown to rest on some prior comprehension of the whole. They are not made of whole cloth from strong-microreductionist investigation.

And negatively, when the implications, effects, and authenticity of the strong-microreductionist perspective *are* carefully examined, they cannot be justified. By removing life from the alive, the strong-microreductionist principle makes the fundamental question of biology—what it is that makes something living—appear naïve and uninformed. It simultaneously impoverishes, compromises, and confuses our research results and, consequently, our understanding of living systems. Even when it seems to offer broad understanding, whether we are considering the mechanism of muscle contraction or of protein secretion, on inspection and in actuality it allows us to confront life only in the most superficial terms.

This situation is somewhat analogous to the development of psychology as an experimental science during the twentieth century. Its emergence was animated by the desire to understand the great mysteries of the mind that Descartes had put beyond science. Lamentably, in the search for this knowledge, in step after insistent step, the purview of the

The Call to Authority

experimental psychologist became increasingly myopic and constricted. The wonders of emotion, thought, self, and intention were seen either as philosophical issues too clouded by metaphysics to be explored by clear-headed scientists or simply as vaporous illusions. For all practical purposes, the discipline came to exclude its own central questions. Mental activity was reduced to the trivial and easily understood. Psychology became scientific and rigorous all right, but in the process lost its way and its mission, willingly locking itself in a Skinner box of simplistic behaviorism, with its all-encompassing, fully explanatory inputs and outputs. The study of the human mind became the study of the human as mindless organic computer.

Like the relationship between behaviorism and mind, the strong-microreductionist principle not only is inadequate to describe life, but severely limits and circumscribes the research carried out under its banner, furnishing a false sense of the reach of our knowledge. I have explained how this works. Deductive testing is replaced by inferences of the weak inductive kind. Consequently, hypotheses and proof often become indistinguishable, even interchangeable. And when this happens, however much we might wish to avoid acknowledging it, there is little choice but to depend on authority for wisdom. As the distinction between model and nature disappears, a favored interpretation can become a matter of opinion masquerading as fact.

Scientists do not like to talk much about this. Whether this is because they honestly believe that today's standing authority—being composed of bright, accomplished, careful, precise, and for the most part unbiased individuals imbued with an uncommon common sense—is perfectly acceptable, or whether their motives are less self-centered and more self-serving, philosophical introspection about the process of

Lessons from the Living Cell

knowing often seems a waste of precious time. Experimental scientists tend more to be doers than reflective thinkers.

In the modern day, when they are put on the defensive by forces that may be neither benign nor informed, scientists unfortunately too often fall back on their own authority to justify their pursuits and perspective. In a recent article a renowned Nobel laureate, Max Perutz, was defending no less a *frère* than Louis Pasteur against a critic of science. The critic argued that in spite of the fact that certain of Pasteur's own observations falsified one of his important hypotheses, Pasteur nonetheless continued to retain his belief in its correctness. The critic saw this as disingenuous at best, and perhaps even dishonest. As we have discussed, holding on to a seemingly failed hypothesis is not uncommon in science. In defending Pasteur, Perutz rightly pointed out that it is foolish to think that a single false observation should necessarily lead to the rejection of an hypothesis or theory. But he then went on to argue that this was not a problem, because we could confidently rely on the scientist's common sense and expert knowledge to determine when to hold on to an hypothesis or theory, and when to abandon it.

This, rather than admitting to the uncertainty of the situation, was an unabashed call to authority. Modern science emerged as a challenge to the common-sense beliefs and expert knowledge of the time. Whether we are talking about the sun traveling around the earth or about life's origins, again and again science has found common sense and expert knowledge, even that of the most knowledgeable people, wanting. The history of science is littered with such views being replaced by counterintuitive but more accurate descriptions of reality. Perhaps today's common sense, or at least the common sense of the modern scientist with his or

The Call to Authority

her special expertise, is nonpareil, beyond failure. But looking at our forebears, this seems unlikely.

Certainly, I do not mean that we should ignore common sense, or the opinions of experts or of a community of experts. But if this is how we justify our beliefs, then we should be clear that they are beliefs of faith or fealty, not obedience to true understanding. And that is why there is silence on the subject. Whatever this problem may be for science in general, the strong microreductionists have no alternative. They *must* rely on the authority of the expert, or whoever is deemed the authority. If what is theory and what is proof is a matter of opinion, as it necessarily is for strong microreductionists, then making the distinction between the two depends on whose opinion matters.

The call to authority is sometimes dressed in the garb of scientific skepticism. It may seem as if scientists are simply demanding strong evidence before being willing to accept a novel theory or hypothesis. On closer inspection, they may actually be making unfair demands of a theory that does not comport with the common view of the time, with current perceptions of nature, which themselves may be viewed more leniently. The claim that rigor is what is being sought is captured by a catchy phrase of the late astronomer and popularizer of science, Carl Sagan. Sagan wrote that "Extraordinary claims require extraordinary evidence." This seems sensible enough on the surface, but is it true? Don't all scientific claims, whether extraordinary or just plain ordinary, require exactly the same quality and character of evidence? Why is the character of the *claim* relevant to the character of the *evidence*? On what logical basis do we require more powerful evidence for one hypothesis than another, even if one claim is seen as being extraordinary? Many of our most established and cherished concepts have at one time or another been thought to be extraordinary hypotheses. Newtonians or Cartesians would have seen relativity, quantum theory, and the genetic code, to name but a few significant examples, as astounding, unbelievable, and most assuredly extraordinary proposals.

Claims are deemed extraordinary by the zeitgeist—the context of the times. Ordinary claims comport well with the various understandings, assumptions, and axioms prevalent at a particular period of time, while extraordinary claims do not. But what is extraordinary today may become ordinary tomorrow—and, significantly, because of very ordinary evidence. For example, the hypothesis that there are little green humanoids on Mars would be viewed as an extraordinary proposal, not merely because it is not consistent with our knowledge of the environment on Mars, but because we have no basis to assume that if life were found there, that it would be in the form of little green humanoids.

Sagan would ask for extraordinary evidence for such an apparently ungrounded hypothesis, and one can certainly understand why. But is it extraordinary evidence that is really required? Let us say that a Martian rover took pictures of little green humanoids, and we followed up with direct face-to-face contact, as with the two convicts at Leavenworth discussed in Chap. 11. As evidence, this seems pretty ordinary—photographs and visual sightings. By obtaining it, the extraordinary hypothesis about little green humanoids would have become quite ordinary. We would know that they exist on Mars, and whatever implications their existence might have for our greater understanding of nature would become part and parcel of our ordinary intelligence. What Sagan was doing, knowingly or not, was tilting the playing field against the uncommon view. His plea was to authority and the common perception, because only authority can label a hypothesis extraordinary, and only authority can judge whether particular proof meets the standard of extraordinary evidence.

Last Word

But authority should not determine the truth or falsity of a theory in science. That should be reserved for the evidence, and scientific evidence is by its very nature quite ordinary in character. In the final analysis, no supposition, assumption, conviction, or strongly held belief can take the place of nature itself, in its wholeness and in its entirety. It is the final authority and arbiter. And to testify to this—to testify to our devotion to *nature's* authority—biology, and science more generally, must test its theories and hypotheses against nature's actual properties, searching those properties over and over again for a more grounded understanding of life and the natural order.

The strong-microreductionists' belief that such a search is unnecessary is verifiably false. To conclude that life can be accounted for, fully entailed, fully understood, as the mere sum of its parts, is to ignore its essential and transcendent quality. And to apply this false belief in our research work, hoping to comprehend the mechanisms that underlie a functioning whole by reasoning solely from our discernment of its presumptive parts, is a counterfeit hope that can lead only to misperceptions, confusion, and dogmatic posturing.

Notes

Chapters 1–3

There are many recent books and articles about the Genome Project and genomics, and it is hard to choose among them. They all tell essentially the same story. A recent volume by the science writer Matt Ridley, entitled *Genome* (Harper-Collins, 2000) is relatively easy to read and touches most of the bases. The perspective of the genetic determinist is most clearly presented in the writings of Richard Dawkins, starting in 1976 with *The Selfish Gene* (Oxford University Press, 1976) [also see, for example, *The Blind Watchmaker* (Longman, 1986) and *Climbing Mount Improbable* (W. W. Norton, 1996)]. Two new books from Harvard University Press raise questions about this perspective [see R. Lewontin, *The Third Helix* (2000) and Evelyn Fox Keller, *The Century of the Gene* (2000)]. Lewontin's book presents a point of view quite similar to the author's, but from a very different perspective.

Chapter 4

For Greek thought on reductionism see C. Bailey, *The Greek Atomists and Epicurus* (The Clarendon Press, 1928), and J. Burnet, *Early Greek Philosophy* (A & C Black, 1892). Hall's *History of General Physiology* (University of Chicago, 1969) provides a valuable outline of the subject from ancient times to about 1900. Newton's laws of motion can be found in his *Principia,* and Descartes's thoughts about life and nature in his various philosophical works [Adam and Tannery, eds., *Descartes' Works* (1897)]. The diversity of thought today about the meaning of the term *reductionism* is expressed in the beliefs of some well-known modern thinkers in J. Cornwell, ed., *Nature's Imagination* (Oxford University Press, 1995).

Chapter 5

In this chapter, the idea of biology as molecule and reaction is confronted by a holistic view of life exemplified by Darwin's theory of evolution [Darwin, *The Origin of the Species by Means of Natural Selection* (1859); Wallace, "On the Tendency of Varieties to Depart Indefinitely from the Original Type," *Proc. Linnean Soc.*, 3:53–62, 1859]. In an important book, *Life Itself* (Columbia University Press, New York, 1991), Robert Rosen draws a conclusion about the nature of life that is similar to that presented here. A good place to start to consider life's origin is the classic book by A. I. Oparin on *The Origin of Life* (Macmillan, New York, 1938) and a useful summary of more recent thoughts can be found in R. Shapiro, *Origins* (Summit Books, New York, 1986).

Chapter 6

In addition to the two examples of the strong-microreductionist approach to the study of living systems discussed in this book (muscle contraction and protein secretion), similar disquisitions can be developed for many fields of study in experimental biology. Research that relies in its genesis, execution, and interpretation on the strong microreductionist principle can be found in a great variety of work published during the past half century in many disparate journals from the clinical *New England Journal of Medicine* or

Notes

Journal of Experimental Medicine, to basic science journals such as *Cell Biology, Cell,* and the *Journal of Biological Chemistry,* as well as in general interest magazines such as *Science* and *Nature.* It is a good exercise for students of biology to see if they can identify this type of research. The strong microreductionist's view of cellular life is well presented in an extensive textbook-level description in B. Alberts et al., *Molecular Biology of the Cell* (Garland Publishing, New York, 2000).

Chapters 7 and 8

Szent-Györgyi's isolation of the proteins actin and myosin was built on the earlier work of A. Danielewsky (*Z. Physiol. Chem.*, 5:162, 1881, 5:349, 1881, 7:124, 1882), who was the first to extract myosin, and W. Halliburton (*J. Physiol.*, 8:133, 1887), who was the first to extract actin. Much of Szent-Györgyi's original research was carried out on incompletely characterized systems, not pure preparations of actin and myosin, most famously, a complex cellular extract called "glycerinated muscle fibers." A great deal of research has been carried out since by many investigators on various extracts of muscle tissue that contain actin, myosin, and a variety of other proteins in attempts to better understand the contractile mechanism (e.g., see the early work of S. S. Spicer and J. Gergely, *J. Biol. Chem.*, 188:179, 1951). Szent-Györgyi summarized his ideas as well as those of others in *Acta Physiologica Scandinavia,* 9: suppl. 25, and *Chemistry of Muscular Contraction* (2d ed., Academic Press, New York, 1951).

Heilbrunn's experiment injecting the calcium ion into whole muscle cells can be found in L. Heilbrunn and F. Wiercinski, *J. Cellular Compar. Physiol.*, 29:15, 1947. Earlier work demonstrated that cutting the ends off of a muscle fiber to expose its interior and then placing it in a calcium-containing medium, caused it to contract. Winograd's autoradiographic studies showing calcium at the cross-bridges can be found in *J. Gen. Physiol.*, 48:455 and 48:997, 1965.

The sliding filament theory of muscle contraction is outlined in most physiology and cellular biology textbooks and was simultaneously proposed by H. E. Huxley and J. Hanson (*Nature*, 173:973, 1954) and A. F. Huxley and R. Niedergerke (*Nature*, 173:971, 1954). Also see H. E. Huxley and J. Hanson, in G. H. Bourne, ed., *The Structure and Function of Muscle*, vol. 1 (Academic Press, 1960) and H. E. Huxley, *J. Mol. Biol.*, 37:507–520, 1968. For a more recent treatment of the subject of muscle contraction, see C. R. Bagshaw, *Muscle Contraction* (Chapman and Hall, New York, 1993) or R. D. Keynes and D. J. Aidley, *Nerve and Muscle* (Cambridge University Press, 1991).

Understanding how electrical impulses are carried along the neuron was in great part the result of work by A. L. Hodgkin and A. F. Huxley (Hodgkin, *The Conduction of the Nerve Impulse*, Thomas, Springfield, IL, 1964); R. R. Kandel et al., *Principles of Neural Science* (3d ed., Elsevier, New York, 1991).

Chapters 9 and 10

The vesicle theory had its origin in Rudolf Heidenhain's nineteenth-century discoveries (Heidenhain, "Beitrage zur Kenntnis der Pancreas," *Pfluger's Arch.*, 10:557–632, 1875), and the similar, and roughly contemporaneous work of W. Kuhne and A. Lea (*Verhandl. Naturhist. Med. Var. Heidelberg*, 1:445, 1876 and *Untersuch. Physiol. Instit. Heidelberg*, 2:448, 1882). An early articulation of the modern vesicle theory is found in G. Palade, P. Siekevitz, and L. Caro,

"Structure, Chemistry and Function of the Pancreatic Exocrine Cell," in A. deReuck and M. Cameron, eds., *Ciba Foundation Symposium on the Exocrine Pancreas: Normal and Abnormal Functions* (Churchill LTD, London, 1962). The original in vivo experiments of Siekevitz, Palade, and Caro were published in a series of articles in the *J. Biophys. Biochem. Cytol.* from 1956–1960 (in vols. 2–7). In the mid to late 1960s, James Jamieson and George Palade carried out similar experiments on pancreatic tissue in vitro and published their results in a series of papers in the *J. Cell Biology* between 1967 and 1971 (see vols. 34, 39, 50).

Subsequent to being awarded the Nobel prize in 1974, Professor Palade outlined his theory and his view of the evidence in *Science* magazine (G. Palade, "Intracellular Aspects of the Process of Protein Synthesis," *Science*, 189:347–358, 1975). Accompanying his article, Keith Porter, a distinguished electron microscopist, wrote of Dr. Palade's accomplishments.

Research that used antibodies linked to gold beads to identify the location of particular digestive enzymes in the acinar cell are presented in J. Kraehenbuhl et al., "Immunocytochemical Localization of Secretory Proteins in Bovine Pancreatic Exocrine Cells," *J. Cell Biology,* 76:406–423, 1977, and M. Bendayan et al., "Quantitative Immunocytochemical Location of Pancreatic Secretory Proteins in Subcellular Compartments of the Rat Acinar Cell," *J. Histochem. Cytochem.,* 28:149–160, 1980.

The notion that new protein enters the cisternal spaces of the endoplasmic reticulum received significant affirmation in studies with isolated microsomes by C. Redman et al. *(J. Biol. Chem.,* 241:1150, 1966). These experiments provided the observational foundation for G. Blobel's and B. Dobberstein's signal hypothesis a few years later *(J. Cell Biol.,* 67:835, 1975a and b).

For a critical evaluation of the vesicle theory and the signal hypothesis see S. Rothman, "Protein Transport by the Pancreas," *Science,* 189:347–358, 1975; S. Rothman, *Protein Secretion: A Critical Analysis of the Vesicle Model* (John Wiley & Sons, New York, 1985); S. Rothman et al., *Nonvesicular Transport* (John Wiley & Sons, 1985); and more recently S. Rothman, "The Sorting of Proteins," in E. Bittar, ed., *Principles of Medical Biology,* vol. 7A: *Membranes and Cell Signaling* (Elsevier, Hampton Hill, Middlesex, U.K, 1997); and L. Isenman, C. Liebow, and S. Rothman, "Protein Transport across Membranes: A Paradigm in Transition," *Biochim. Biophys. Acta,* 1241:341–370, 1995.

The use of inductive inferences in modern science is usually traced to Francis Bacon's influential writings in the early 1600s. His ideas can be found in the *Advancement of Learning* (1605) and *New Organum* (1620). Inductive reasoning as a scientific method has been under attack primarily by philosophers for most of the twentieth century. For an in-depth analysis of the problem with induction, as well as the centrality of the deductive testing of theories see K. Popper, *The Logic of Scientific Discovery* (Harper and Row, New York, 1965).

Chapters 11 and 12

The "nonparallel" secretion of various digestive enzymes was first reported in Rothman, "Non-Parallel Transport of Enzyme Protein by the Pancreas," *Nature,* 213:460–462, 1967. This phenomenon and its implications have been reviewed in Rothman, "The Digestive Enzymes of the Pancreas: A Mixture of Inconstant Proportions," *Annual Review of Physiology,* 39:373–389, 1976; S. Rothman, "The

Notes

Regulation of Digestive Reactions by the Pancreas," in J. G. Forte, ed., *Handbook of Physiology: The Gastrointestinal System,* vol. III (Oxford University Press, New York, 1989), pp. 465–476; and in S. Rothman, C. Liebow, and J. Grendell, "Nonparallel Transport and Mechanisms of Secretion," *Biochim. Biophys. Acta,* 1071:159–173, 1991.

Membrane transport for proteins was first reported by C. Liebow and the author (C. Liebow and S. Rothman, "Membrane Transport of Proteins," *Nature,* 240:176–178, 1972) and much of the early work on the pancreas is reviewed in S. Rothman, "The Passage of Protein through Membranes: Old Assumptions and New Perspectives," *Amer. J. Physiol.,* 238:G391–G402, 1980; S. Rothman and C. Liebow, "The Permeability of the Zymogen Granule Membrane to Digestive Enzymes," *Amer. J. Physiol.,* 248:G385–G392, 1985; and S. Rothman, *Protein Secretion: A Critical Analysis of the Vesicle Model* (John Wiley & Sons, New York, 1985).

The large body of work describing membrane transport for proteins over the last decade or two has been reviewed in L. Isenman, C. Liebow, and S. Rothman, "Protein Transport across Membranes: A Paradigm in Transition," *Biochim. Biophys. Acta* 1241:341–370, 1995, and in various articles in a series of three monographs entitled *Membrane Protein Transport* (S. Rothman, ed., Elsevier, vols. 1–3, 1995–1996).

Work in the 1990s by K. Goncz, microscopically observing and quantifying protein transport across the membrane of single secretion granules, can be found in K. Goncz and S. Rothman, "Protein Flux across the Membrane of Single Secretion Granules," *Biochim. Biophys. Acta,* 1109:7–16, 1992; K. Goncz and S. Rothman, "A Transmembrane Pore Can Account for Protein Movement across Zymogen Granule Membranes," *Biochim. Biophys. Acta,* 1238:9193, 1995.

Chapter 13

The social context of the scientific enterprise is described in many writings, but first among them is Thomas Kuhn's *The Structure of Scientific Revolutions* (University of Chicago, 1962).

Index

A

Acinar cells, 139–146, 149, 228, 241–242, 270–271
Actin, 106, 122, 125, 128, 129
Actin-myosin theory of muscle contraction, 106, 115, 122, 125, 129
Adaptation, 78–89
  See also Evolution
Adenosine triphosphate (ATP), 106, 123, 132
Aggregation, molecular, 61–62, 70–74
Allen, Robert Day, 167
Amino acids, 69, 178–179, 187–189, 193, 225–226, 238, 263–264
Amphiphilic molecules, 263
Amylase, 223, 225
Aristotle, 15
ATP (*see* Adenosine triphosphate)
Autoradiography, 122–123, 180, 193–208, 211–212

B

Babkin, B. P., 223–227
Bernard, Claude, 11
Bertillon, Alphonse, 218
Bertillon method, 218–219
Bilayers (of lipid membranes), 262–263, 265–267, 270
Biochemical reactions, 223
Biological evolution, 84–85
Birds, 86
Blobel, Gunther, 268
Brookhaven National Laboratory, 261

C

Calcium, 113–116, 122–123, 133
cAMP (cyclic adenosine monophosphate), 229–230
Capacitance, electrical, 168

Cartesianism, 24–25, 137, 165
CCK (cholecystokinin-pancreozymin), 244
Cell biology, 136, 142–143, 275
Cell division, 13
Cell fractionation, 179–182, 184–186, 189, 193–194, 211–212, 244, 245
Cell membranes, 59, 71, 151, 152, 160, 250–251
  electrical capacitance of, 168
  and exocytosis, 227
  permeability of, 234, 236–239
  polar aspects of, 262–265
  transport mechanisms of, 233–240, 268–270
Cell studies, 110, 126, 142–143
Cell theory, 12–14, 110
Cells, 61–64, 113–114
Centrifugation, 179, 181, 188
Cholecystokinin-pancreozymin (CCK), 244
Chymotrypsinogen, 223, 225–226, 244
Cis (Golgi saccules), 152
Cisterna, 269
Cisternal spaces, 151, 152
Cohen, Leonard, ix
Coloration, protective, 85–86
Compartmentalization, 56–57, 209–210
Competitive inhibition, 249
Condensing granules, 151
Condensing vacuoles, 152, 182, 187
Confocal imaging, 167
Conservative electron microscopists, 143–144
Constitutive pathway, 250
Cortex, cell, 113, 114, 123
Cotranslational translocation, 188, 193

Index

Criminal identification, 218–220
Cross-bridges, 121–123, 127
Cyclic adenosine monophosphate (cAMP), 229–230
Cytoplasm, 57, 58, 111–118, 128–129, 145, 185–187, 189, 192, 209–214, 242–246, 257, 259, 267, 268

D

Darwin, Charles, 4, 7, 79–80, 82, 84, 86
Deductive logic, 161–163, 173, 281
Deductive testing, 221
Democritus, 22, 23
Deoxyribonucleic acid (*see* DNA)
Depolarization, 123
Descartes, René, 15, 22–23, 47, 62, 241, 280
Descartes's machines, 61, 62
Determinism, 37–39
Diffusion, 234–239, 255
Diffusion-based membrane transport, 242
Diffusional equilibria, laws of, 255
Diffusional model of protein transport, 234–239
Digestion:
 Pavlov's/Babkin's studies of, 223–225
 Rothman's studies of, 225–226
 See also Enzymes, digestive
Direct transport, xvi, 233–240, 266
Disorder, 68, 83
DNA (deoxyribonucleic acid), 2–7, 43, 46, 57, 60, 64–66, 70–71, 77, 78, 81, 90–91, 94, 96, 182
Dynamic universalism, 30–31

E

Einstein, Albert, 28, 217
Electrical capacitance, 168
Electrical impulses, 130, 166
Electrical stimulus, 123
Electricity, Galvani's study of, 103–104
Electron microscopes, 117–122, 180–181
 autoradiography cell studies, 111–112, 136
 and function/structure argument, 171–172
 muscle contraction, study of, 127, 132
 pancreas, studies of, 141–145
 and structures, 148–149
 and vesicle theory, 150–151
 See also Autoradiography
Electron microscopists, 142–144, 166, 178
Endogenous proteins, 153
Endoplasmic reticulum (ER), 59, 145–152, 154, 159, 169, 179, 181, 182, 187–192, 212–214, 267–269
Endothelilal cells, 170
Entropy, 68, 83
Enzymes, digestive, 139–141, 144–145, 153, 154, 180–182, 185–186, 188, 189, 192, 213–214, 228–229, 242–244, 251–252, 249, 258, 268, 270–271
Epinephrine, 87
ER (*see* Endoplasmic reticula)
Eucaryotic cells, 138, 144, 145
Evolution, 4, 8, 68, 77, 79–91
Exocytosis, 152, 156–157, 159–160, 162–163, 167, 168, 171, 173, 227–232
Exogenous proteins, 153
Experimental biology, 7, 94–100
Experimental protocols, 179–180

F

Fight-or-flight reaction, 87–88
Filaments, 121–122
Finches, 86
Fingerprinting, 218–221
First messengers, 229–230
Fossil record, 67, 69–70
"Free will," 89
Functional models, 147–150

Index

G
Galápagos Islands, 86
Galen, 102–104, 124
Galvani, Luigi, 103–104
Gel state, 113
Gene therapy, 64
Generalization, 25–26
Genes, 4, 8, 64, 96–97
Genetic code, 2, 7, 80
Genetic determinism, 4, 5, 8
Genetic recombination, 7
Genomic Age, 8
Genomics, 1, 3–4
Geological evolution, 82–84
Gödel, Kurt, 28, 29, 41
Goldberg, Rube, 137
Golgi apparatus, 145–150, 152, 154, 169, 182, 183, 187, 212–213
Golgi saccules, 152
Grand universalism, 27–29
Granule lysis, 260
Granules, 154, 160, 167–169, 229, 230, 257–259

H
Heart muscle, 113, 131–132
Hegel, Georg, 124
Heidenhain, Rudolf, 139–142, 151, 167
Heilbrunn, Lewis, 167
  and calcium studies, 113–116, 122–123, 133
  and cell studies, 112–118
  and protoplasm, 109–110
  and sol/gel transformation, 128–129
  and study of muscle contraction, 107–110, 114–118, 120–124, 128–129
  and Szent-Györgyi, 115–118, 133–134
Hierarchical universalism, 26–27
Hilbert, David, 41
Hill, A. V., 105, 120
Hippocratic oath, 275
Histamine, 167
Homogenization, 170, 244–246

and granule permeability, 258
and microsomes, 187, 191
and pancreatic studies, 179, 181–185, 187
and zymogen granules, 251–252
Hooke, Robert, 12, 109
Human body, 22–23
Human Genome Project, 3–4, 96, 172
Huxley, Hugh, 121, 122, 129, 132

I
I band (isotropic), 120–121
Incompleteness theorem, 28, 41
Inductive reasoning, 159–163, 172, 276, 281
Interstitial spaces, 170
Inulin (NB), 191
Isotropic (I) band, 120–121

J
Jamieson, James, 178

K
Keats, John, 197
Kuhn, Thomas, x–xii, 135

L
Lack, David, 86
Laws of diffusional equilibria, 255
Leavenworth (Kansas), 218
Life, 2–3, 5–9, 13–15, 43, 46–47, 66–73, 78–80, 82, 89–91, 94, 110
Linear polymers, 75
Lipid bilayer model, 262–263, 265–267
Lipid membranes, fragmentation of, 169
Lipids, 71, 237–239
Lysis, 260
Lysosomes, 59, 153

M
Machines, 22–24, 61–64, 130
Mannitol, 238
Mast cells, 167
Material substance, 101

Index

Materialism, 37
Matter, 83, 90–91
McGill University, 224
Mechanisms, xiv, 129–132, 135–138
Medial (Golgi saccules), 152
Medulla, cell, 113–114, 123
Membrane transport, 233–240, 248–271, 274
Membranes, 57, 58, 169, 238–242, 251–252, 256, 260, 262, 266
Microfibrils, 75
Microfilaments, 57
Microreductionism, 39, 117–118, 127, 244–245, 258
  See also Strong microreductionism; Weak microreductionism
Microsomes, 181, 182, 187–192, 259
  and granule permeability, 259
Microtubules, 57, 75
Microvesicles, 146, 170, 182
Mitochondrion, 57–59, 75, 93, 145, 209–210, 259, 268
Molecular transport, 147, 153, 168–170, 233–240, 251–271
Molecules, 5–6, 8, 43, 61–62, 70–74, 107, 189, 192, 238
  amphiphilic, 263
  and strong microreductionism, 93
Muscles, 101–118, 120–129, 131–134, 144
Mutations, 7, 78
Myosin, 106, 122, 125, 128, 129

N

Naïve inductivism, 128
National Institutes of Health (NIH), 51
Natural law, 24, 30
Natural selection, 4, 6–9, 78–84
Nerve synapse, 166
Neurons, 130–131
Neurotransmitters, 136, 137, 141, 154
Newtonian physics, 30
Newton's laws, 22, 23, 31, 38

Niebuhr, Reinhold, 273
NIH (National Institutes of Health), 51
Nonparallel secretion, 226, 229–231, 241–271
Nucleic acids, 70, 71
Nucleus, 58, 59, 93
Numbers theory, 28–29

O

Objective observation, 35–37
Objectivism, 35–37
Ockham, Sir William of, 33
Ockham's razor of parsimony, 33–35
Oil-soluble substances, 237, 239
Omega figures, 151, 156–157, 159–160, 162–163, 171
Ordered arrays, 57
Ordered environments, 57–59
Origin of life, 66–77
Oxidative energy metabolism, 75

P

Palade, George, 178, 181
Pancreas, 139–143, 165–166, 251, 257, 271
  Pavlov's theory of, 223
  Rothman's study of, 225–229
Paradigms, xi–xii
Parallel secretion, 224, 226, 227, 230–231
Parsimony, 33–35, 228, 234–236, 242
Pasteur, Louis, 67, 282
Pavlov, Ivan, 223–226
Peptide bonds, 225–226
Peptide chains, 188, 264–265, 269
Peptide hormones, 137, 141, 154, 226
Permeability, 234, 236–239, 241–242, 251–252, 256–259, 261, 266, 270–271
Peroxisomes, 59
Perutz, Max, 282
Planar organization, 75
Planck, Max, 278
Planck's law, 278

Index

Plasma proteins, 170
Polysaccharides, 191
Popper, Karl, 221–222, 278
Progressive electron microscopists, 144
Protective coloration, 85–86
Protein synthesis, 145, 181, 267, 270
Proteins, 2–3, 57, 62, 65, 66, 69–71, 87, 110, 111, 138, 139, 151–153, 156, 178–180, 183–193, 209–213, 233–240, 242, 247–271
  and acinar cell studies, 145–146
  and Human Genome Project, 172
  and muscle contraction, 106, 115, 122, 123, 128, 129
  Pavlov's/Babkin's studies of, 223–225
  and signal hypothesis, 268–270
  and space/time constraints, 74, 75
  and strong microreductionism, 43, 80, 81
Proteolytic enzymes, 225
Protoplasm, 109–111
Protoplasmic theory, 111, 137, 142
Psychology, 280–281

R

Radioactive isotopes, 194
Reductionism, xiii, 12, 19–44
  See also Microreductionism; Strong microreductionism; Weak microreductionism
Reproducibility, 36
Resolution (autoradiography), 202–205
Ribonucleic acid (*see* RNA)
Ribosomes, 59, 145–146, 150–153, 179, 181, 183, 188–190, 192, 209, 214, 249, 267, 269
Ringer, Sidney, 113
RNA (ribonucleic acid), 5, 57, 71, 77, 145, 181, 188, 267
Rockefeller Institute, 178

S

Saccules, 150, 152
Sagan, Carl, 283, 284
Sampling (autoradiography), 199–202
Sarcoplasmic reticulum, 123
Schlieden, Mathias, 12
Schwann, Johannes, 12
Science magazine, 1, 166
"The Search for Truth" (René Descartes), 47
Second messengers, 229–230
Secretion, parallel/nonparallel, 226
Secretion granules, 59, 140–141
Secretion mechanism, 226–233
Secretion systems, 250
Sedimentation, 179, 181–183
Self-replication, 71
Sequestration, 189, 192, 214
Serotonin, 167
Siekevitz, Philip, 178
Signal hypothesis, xviii, 138, 151–152, 268–270
Skeletal muscles, 120
Sliding filament model, 121–122, 127, 128, 132, 139
Sol state, 113
Sol/gel theory of motion, 113–114
Sol/gel transformation theory of muscle contraction, 115, 128–129
Solubility of proteins, 239
Solvent water, 72
Space constraints, 56–59, 73–77, 82
Spatial arrays, 75
Spatial locations, 56–59
Specific radioactivity, 258
Spontaneous generation of life, 67, 73
Starch, 223
Static universalism, 31–32
Steady state, 255
Steroids, 87
Stimulus-response coupling, 133
Striated muscles, 120, 131
Strong microreductionism, 40–44, 48, 86, 93–101, 278–280
  and adaptation, 78–80

Index

Strong microreductionism *(cont)*
  and anatomical fidelity, 150
  and authority, 283
  and biological research, 42–43
  and cell theory, 47, 60, 63
  compartmentalization, 210
  and inductive reasoning, 172
  and muscle studies, 125–126
  and naïve inductivism, 128
  and progressive electron microscopists, 144
  and psychology, 281
  and Szent-Györgyi, 104–107
  and vesicle theory, 135–137, 158–164, 169, 171, 172, 177, 178, 215, 273, 276–278
Strong static universalism, 32–33
Structure, function vs., 143–144
The Structure of Scientific Revolutions (Thomas Kuhn), x–xii
Superprecipitation, 106, 122, 126, 129, 138
Syncytia, 13
Szent-Györgyi, Albert, 104–109, 115–118, 120–126, 129, 133–134, 138

T
Targeted sites, 57
Thermodynamics, second law of, 68
Third messengers, 230
Time constraints, 59–60, 76–77, 82
Tinguely, Jean, 137
Trans (Golgi saccules), 152
Trans-Golgi network (Golgi saccules), 152
Transgenerational entailment, 66
Transgenic animals, 173
Transport, vectorial, 257
Transport process, 176, 177, 189–190
Tritium, 180
Trypsinogen, 223, 225–226, 244

U
Universalism, 24–33
University of Chicago, 109

University of Pennsylvania, 107, 133, 225

V
Vacuoles, 151
Vectorial transport, 257
Vesicle mechanisms, 242
Vesicle theory of protein secretion, xv–xviii, 150–166, 168–172, 175–215, 223–229, 232–233, 245–248, 271
  and cell theory, 139
  criticism of, 274–275
  direct transport vs., 234–240
  paradox of, 268
  and strong microreductionism, 135–137, 273, 276–278
Vesicles, 146–149, 151, 153, 168–170, 267
Visible-light microscope(s), 117, 143, 145, 167, 173
Vitalism, 20–21

W
Wallace, Alfred, 4
Water-soluble substances, 237, 239
Wavelengths, 261
Weak microreductionism, 39–40, 94
  and biological research, 42
  and experimental biology, 97
  Heilbrunn and, 108
West, William, 218–220
Wiercinski, Floyd, 114–116, 129, 134
Winograd, Saul, 122
Wolfe, Alan, 175
Woods Hole Marine Biological Laboratory, 105, 115, 133–134

Z
Z lines, 120, 121
Zymogen granules, 140–142, 145–148, 150–154, 179, 181–183, 185, 187, 212, 242, 245, 249–262